Hunger and Postcolonial Writing

Hunger and Postcolonial Writing explores contemporary postcolonial fiction and life-writing from various geopolitical contexts.

The focus of this work is hunger – individuated in the self-imposed starvation of the hunger protester, and on a mass scale in the form of famine and food insecurity. It considers the hungry colonial and postcolonial body, examines its textual forms and historical trajectories, and situates it within the food security context of imperialism and its legacies. This book is the first monograph-length study of hunger within a postcolonial/world literary context. Its transcolonial focus produces comparative readings across postcolonial writings, facilitating productive analyses of the operations of imperialism and its after-effects across heterogeneous zones of colonialism. This project reads hunger as defined by the social, cultural, historical, and economic engagements produced by colonial and postcolonial encounters. Examining the starving colonialized body through Cartesian models of somatic subjectivity and considering how this body is mediated by post-Enlightenment discourses of Modernity and progress, this work interrogates the contradictions produced by the starving colonial body as it is positioned between the possibility of radical protest and prescriptive colonial discourse.

This book will appeal to scholars and students interested in food studies, postcolonial studies, human geography, sociology, and literature studies.

Muzna Rahman is a lecturer in English at Manchester Metropolitan University, UK.

Critical Food Studies
Series editors: **Michael K. Goodman**, *University of Reading, UK*, and **Colin Sage**, *Independent Scholar*

The study of food has seldom been more pressing or prescient. From the intensifying globalisation of food, a world-wide food crisis and the continuing inequalities of its production and consumption, to food's exploding media presence, and its growing re-connections to places and people through 'alternative food movements', this series promotes critical explorations of contemporary food cultures and politics. Building on previous but disparate scholarship, its overall aims are to develop innovative and theoretical lenses and empirical material in order to contribute to – but also begin to more fully delineate – the confines and confluences of an agenda of critical food research and writing.

Of particular concern are original theoretical and empirical treatments of the materialisations of food politics, meanings and representations, the shifting political economies and ecologies of food production and consumption and the growing transgressions between alternative and corporatist food networks.

Food System Transformations
Social Movements, Local Economies, Collaborative Networks
Edited by Cordula Kropp, Irene Antoni-Komar and Colin Sage

Metaphor, Sustainability, Transformation
Transdisciplinary Perspectives
Edited by Ian Hughes, Edmond Byrne, Gerard Mullally and Colin Sage

Food and Cooking on Early Television in Europe
Impact on Postwar Foodways
Edited by Ana Tominc

Hunger and Postcolonial Writing
Muzna Rahman

Food Sovereignty and Urban Agriculture
Concepts, Politics, and Practice in South Africa
Anne Siebert

For more information about this series, please visit: www.routledge.com/Critical-Food-Studies/book-series/CFS

Hunger and Postcolonial Writing

Muzna Rahman

LONDON AND NEW YORK

First published 2022
by Routledge
2 Park Square, Milton Park, Abingdon, Oxon OX14 4RN

and by Routledge
605 Third Avenue, New York, NY 10158

Routledge is an imprint of the Taylor & Francis Group, an informa business

© 2022 Muzna Rahman

The right of Muzna Rahman to be identified as author of this work has been asserted by her in accordance with sections 77 and 78 of the Copyright, Designs and Patents Act 1988.

All rights reserved. No part of this book may be reprinted or reproduced or utilised in any form or by any electronic, mechanical, or other means, now known or hereafter invented, including photocopying and recording, or in any information storage or retrieval system, without permission in writing from the publishers.

Trademark notice: Product or corporate names may be trademarks or registered trademarks, and are used only for identification and explanation without intent to infringe.

British Library Cataloguing-in-Publication Data
A catalogue record for this book is available from the British Library

Library of Congress Cataloging-in-Publication Data
A catalog record has been requested for this book

ISBN: 978-1-138-69796-6 (hbk)
ISBN: 978-1-032-22140-3 (pbk)
ISBN: 978-1-315-50593-0 (ebk)

DOI: 10.4324/9781315505930

Typeset in Bembo
by Taylor & Francis Books

To Ammu and Bapi, who allowed me to stray so far from home, to find and pave my own path.

Contents

	Acknowledgements	viii
1	Introduction – (Post)colonialism, Hunger, and the Body	1
2	(Post)colonial Foodways and Transhistorical Hungers in Kiran Desai's *The Inheritance of Loss*	35
3	The Text, Starving Body, and J.M. Coetzee's *Life & Times of Michael K*	68
4	Anorexic Fictions and Starving Histories in Tsitsi Dangarembga's *Nervous Conditions*	101
5	Traumatic National Hungers and the Starving Irish Body: Bobby Sands' 1981 Hunger Strike	131
	Index	179

Acknowledgements

Thank you to everyone at Routledge and beyond who supported me with patience and kindness during the process of publishing this work – Faye Leerink, Nonita Saha, Ruth Anderson, Pris Corbett, and Katherine Laidler. A special thanks to Series Editor, Professor Mike Goodman, who made the process as stress-free as possible and was always an encouraging and supportive virtual presence throughout.

I am in debt to my PhD supervisors, Dr Anastasia Valassopoulos and Dr Robert Spencer, who helped me start, shape, and finish my doctoral project upon which this book is based, and provided exemplary models of how to both teach and be taught.

Thank you to all my friends at the University of Manchester, Manchester Metropolitan University, and beyond – too numerous to name – for their constant help, support, and company.

Thank you to my extended family in Bangladesh – especially my seven stalwart Aunties – who gave me care, shelter, and the comfort of family while I began reworking this research for publication.

A special thanks to Dora for supporting me while I pushed this project over the finish line. Support can take many forms, and you provided every kind in abundance.

1 Introduction – (Post)colonialism, Hunger, and the Body

This book examines hunger in contemporary postcolonial writing. Using an interdisciplinary Food Studies theoretical framework, with a particular focus on gastrocritical approaches, this project interrogates the various narrative and political forms produced by the dialectics of self-imposed hunger, by the colonial/postcolonial subject in postcolonial writing. Other forms of hunger are also explored throughout this work, namely instances of food insecurity and famine. The hunger strike is placed within this historical and socio-political context of food deprivation. By situating the hunger-striking body within a larger material history of hunger, this book frames both types of hunger as politicized conditions, entrenched within a larger legacy of British imperialism. Hunger-as-protest takes on a myriad of forms in this study, and this scholarship is largely concerned with the potential political value (or lack thereof) in these forms of protest, as well as a variety of ontological, political, and epistemological contradictions that the hunger strike produces.

This work focuses on contemporary postcolonial writing from 1980 to 2006. The texts are all set between the 1960s and the 1980s. Although there are elements of transnationality in some of these texts, largely they refer to Zimbabwe, South Africa, India, and Northern Ireland. Thus, although geographically broad, the historical focus of this project is relatively narrow. The historical limits represented by these texts, and the common forms of British imperialism considered, enable situating these texts side by side, as do the text's indebtedness into an Anglophone literary tradition. However, the differences between them are widened considerably by the different socio-cultural and historical contexts that each text covers. This project's focus on the specific forms of national and self-imposed hunger, I believe, closes this gap. Hunger 'occupied a central place in both the English and the colonial imagination.'[1] I examine these texts in the spirit of transcoloniality – 'in that it connects heterogenous (post)colonial sites in a critical and comparative exploration of coloniality.'[2] As Françoise Lionnet and Shu-mei Shih argue in *Minor Transnationalism*,[3] 'lateral' comparisons across disparate socio-political sites of colonialism can be productive, especially within the context of globalization and the interconnectedness of nation states. The postcoloniality interrogated in these texts, in Deepika Bahri's words, examines 'different writers taking up a temporally discontinuous but thematically connected screed on

DOI: -1

postcolonial hungers, food, and power.'[4] Indeed, the cultural contexts from which my texts emerge are 'discontinuous' in a variety of ways – J.M. Coetzee's *Life and Times of Michael K* is set in an allegorical South Africa, Bobby Sands' hunger narratives represent more straightforward forms of life-writing in the context of the Troubles, Tsitsi Dangarembga's *Nervous Conditions* is set in the colonial period of Zimbabwe's history, and Kiran Desai's *The Inheritance of Loss* presents a sprawling, globalized postcolonial narrative. However, the universality of eating and hunger in its various forms, and the common forms of European discourses on Modernity, civilization, and colonialism that constitute the matrix of hungers examined in these texts, provides a legible and consistent logic through which to read the starving bodies in this study together.

The food abnegation examined in this study is not easily identifiable as hunger strikes in the conventional sense: that is, overtly politically motivated and clearly declared as an attempt at bodily resistance by the strikers themselves. Excepting the example of Bobby Sands in the fourth and final chapter, these individuals are all fictional characters in novels. Also, apart from Bobby Sands, these hungry bodies are decidedly silent on the matter of interpreting their own hunger. They perform a starvation whose decipherability is left to the reader/viewer to interpret. It is my interpretation that these episodes of hunger are forms of dissent; thus I use the term 'hunger strike' or 'protest' to describe them. I argue that these hungers are a response to the fractured subjectivity resulting from postcolonial/colonial contexts and histories. By tracing individual protesters through their specific socio-political trajectories and then placing them within a larger historical narrative of food insecurity, I explore how hunger protests are not simply a response to current postcolonial/colonial contexts, but also re-articulate a material history of hunger, famine, and lack, focalized through the inequities of power imperialism produces. 'Given the pervasiveness of hunger under conditions of slavery and colonialism, it is entirely fitting that alimentation has assumed a certain pre-eminence as an aesthetics and a politics in anti-colonial critique as well as in postcolonial literature and culture.'[5] Despite their divergent geopolitical and historical contexts, the hunger strikes studied in this work share many common characteristics. A comparative and transnational analysis enables readings that connect various permutations of colonial discourse from the metropole to the specific peripheries of the colonial space.

The texts I examine are situated within a contemporary postcolonial literary tradition but are explorations of bodies under colonialism. Thus, the hunger strikes contained within this study are subject to postcolonial theoretical frameworks, but their hunger strikes may be conceived as anti-colonial struggles and are located and analyzed within specific colonial and historical contexts. Cultural Materialist and/or New Historicist framings are useful here (although the historical period of study is relatively narrow) – that is, reading the text in the context of the critic, situating the text within its own historical setting, and reading literary writing alongside and against historical sources. The contemporary publication dates of the texts themselves necessarily facilitate (post)colonial literary approaches, and some texts examined articulate connections between the colonial and decolonized period in definite ways – considering, for example, globalization and neocolonialism. Thus, it may be helpful to distinguish the postcolonial as it is applied in this work. I deploy the term postcolonial as reaching far beyond the historicization that the 'post' suggests. Although the term postcolonialism has become so heterogeneous and diffuse that a clear definition is almost impossible, one vital framing for the purposes of this project comes from Ania Loomba: 'It has been suggested that it is more helpful to think of postcolonialism not just coming literally after colonialism and signifying its demise, but more flexibly as the contestation of colonial domination and the legacies of colonialism.'[6] Therefore, by postcolonial I do not refer to a historical time period after decolonization within a specific geopolitical space, although this certainly is one definition of the term postcolonial.[7] In this project, I use the term to refer to a constellation of themes and motifs that relate to the construction of racial and national identity of subjects that experience the effects of imperial ideologies – past and present. Within the context of the texts I examine, I refer to a general condition of coloniality and/or postcoloniality.

The starving bodies considered in this book can be read as politicized acts because they are direct retorts to imperial strategies and ideologies, and the characters examined live under the yoke of various forms of colonial rule. Anti-colonial terminology is also useful here if we remind ourselves that: 'The "colonial" still exists and failing to include the anti-colonial in the current neo-colonial moment is very problematic and limiting to intellectual discursive practices that seek liberation and decolonization.'[8] I use the term anti-colonial sparingly, however, as most of the hunger strikes in this book are not self-proclaimed political acts of resistance – and the hunger strikes themselves are better understood as ontological resistances and operate primarily on theoretical and discursive levels. This is not to say the anti-colonial should be disregarded – after all, '[a]s a philosophical movement and critical analytic, anticolonialism is the under-acknowledged predecessor to postcolonial theory'[9] – but the more particular histories of activism and group agitation associated with anti-colonial struggle are not represented in the texts in this study, bar the example of Bobby Sands. I consider the more overt, anti-colonial political goals achieved by Sands' hunger strike in more detail in Chapter 4, as well as interrogating the politicization (or lack thereof) of Coetzee's literary oeuvre in Chapter 2.

This work draws on methodologies outlined in Maud Ellmann's *The Hunger Artists: Starving, Writing and Imprisonment*,[10] as well as Susan Bordo's work, such as *Unbearable Weight: Feminism, Western Culture and the Body*[11] and *The Flight to Objectivity: Essays on Cartesianism and Culture*.[12] I am heavily indebted to Ellmann's scholarship, which connects acts of eating and starving to writing, language, and discourse, as well as her deployment of certain models of the body and subjectivity that I explore and expand upon in this chapter, and throughout the rest of this study. Where my work diverges from Ellmann's is its concerted focus on historicizing and cultural particularism. My readings are situated within the wider context of British imperialism and although my primary materials are sourced from a wide range of geographical spaces, I do not make the ambitious jumps in phenomenological analysis that Ellmann accomplishes so deftly in her work, preferring to situate my work within the historical limits that postcolonial theory typically navigates. I am also greatly inspired by Bordo's work on embodiment, the discipline of the gendered body, and body/mind dualisms. Although I am interested in the gendered aspects of the body under colonialism and read them through the modern gendered Cartesian body – the body/mind dichotomy – I am more focused on how this dichotomy frames the racialized body. The focalization of my work through the lens of imperialism, postcolonialism, and race is what distinguishes my work from Ellmann's and Bordo's excellent scholarship. Given that both critics engage with ideas of biopower and the disciplinary, this book seeks to fill a gap in food scholarship and applies these concepts as they operate within the imperial/postcolonial space. Vitally, I anchor my work within a material history of food insecurity and famine, emphasizing the specific historical and interdisciplinary concerns of this book.

Food Studies and Literature

The study of food may appear at first glance to be a highly specialized field of study within an academic context, but this could not be further from the truth in light of the flurry of scholarly activity this field has inspired in recent years, across a variety of disciplines. In fact, much of the trickiness in discussing food and eating is the sheer scale of investigation that this ubiquitous object and mundane activity has produced. Published in 2018, the fourth edition of *Food and Culture: A Reader*,[13] a consolidation of interdisciplinary food-related analyses, attests to the establishment of food scholarship as a valid and popular investigative concern in the humanities and social sciences.[14] In the 20 years since the release of its first edition in 1997, the field has expanded into multiple and often disparate disciplines – a testament to the rich inter- and trans-disciplinary potential of the topic. The aptly named field of Food Studies has been firmly established, particularly in the social sciences, but aside from its prevalence there, food and eating as a target of investigation has found its expression in cultural studies,[15] history,[16] film studies,[17] philosophy,[18] geography,[19] architecture,[20] archaeology[21] – to name but a few disciplines.

The diversity of disciplinary approaches is matched only by the variety of food-related topics under consideration. Cookery shows, eating disorders, diet books, food security, the slow-food movement, vegetarianism and veganism, body image, restaurant and fast-food culture, personal food narratives, organic food trends, genetically modified foods, food fetishes – the possibilities are seemingly endless. Eating practices, the lack of food, the overabundance of food, the price of food, the look of food, and the meanings of food are all concerns of Food Studies. It is certain that:

> once food became a legitimate topic of scholarly research, its novelty, richness, and scope provided limitless grist for the scholarly mill – as food

links body and soul, self and other, the personal and the political, the material and the symbolic.[22]

It is food's ability to connect the self to others and to situate the individual in relation to community that imbues it with rich scholarly possibility. Highly personal yet inevitably universalizing, food permeates every culture and society and, consequently, comes to bear on every identity, history, and location in the world. The ever-mutating meanings of this mundane but essential cultural object guarantee its importance in popular and scholarly concerns. Put simply, people will always need to eat, and therefore there will always be an essential place for food in the human consciousness, social formations, and various academic arenas.

The field of Food Studies has covered this vast range of topics and, using a variety of methodological techniques, has established certain trends and accepted scholarly norms in its approaches. Foundational texts in the field[23] tend towards the anthropological and sociological; in fact, Food Studies has acquired a certain bias toward social scientific approaches. Elspeth Probyn notes in her book *Carnal Appetites*:

> In its sincere and authentic mode, food writers by and large serve up static social categories and fairly fixed ideas about social relations. It is, however, more worrisome to find the same trends in the field of food sociology, where the establishment of the proper way to study food meets with a certain reification of food as a proper sociological object.[24]

The reification of 'proper study' Probyn mentions is evidenced in the most recent edition of *Food and Culture: A Reader*, revealing its dominant social scientific flavour. Although the text mentions various literary contributions in its introduction, its entire edited collection does not contain a contribution that includes literature as its primary text or field of investigation, although several books have been published along these lines.[25]

Literature remains largely cordoned off from the field of Food Studies.[26] Literary Food Studies – or gastrocriticism, as it is sometimes referred to – is received as a distinct scholarly field, with perhaps different aims and certainly different methodologies to the disciplines most commonly associated with Food Studies. The most obvious distinction between these subjects and literature is that the former claims 'the real world' as their object of study, whereas the latter is primarily concerned with fictionalized narratives. There exists a general understanding that food, eating, and hunger deserve the rigorous, more empirical treatment that subjects like history, sociology, cultural studies, or anthropology are understood to provide. These forms of scholarly engagement emphasize cultural and historical specificity. Additionally, 'useful data' in sociological or anthropological terms tacitly implies that it yields real-world usefulness, such as data that may affect policy. In the case of hunger, this is even more vital. Academic studies of food insecurity are often oriented towards hunger alleviation and attempt to produce positive social change.[27] This understanding provides some clue as to why literature has perhaps been marginalized in the debate on food. Real famines – not imagined representations of them – need to be rigorously studied and understood to formulate ways of overcoming them. Literature is understood to somehow lack the proper register required for these sorts of interventions.

This project considers how intersections between the gastrocritical and more empirical disciplines within Food Studies might generate new and productive insights. I explore how literary interventions can also provide useful analyses – its own 'data' – that can operate productively within the critical register of Food Studies. With its focus on symbolic and mimetic representations, literature provides a useful platform from which to interrogate some of the epistemological norms of Food Studies. But it is the ubiquity of food and the experience of eating (and, as this study is particularly concerned with, hunger) that makes it fertile scholarly ground for inter- and multidisciplinary analysis, focalized through literary studies. It is the universality of food that determines its importance as an area of study, and often the required specificity of food analysis (a specificity that disciplines within Food Studies often deem of the utmost importance) may miss the opportunity to conduct comparative analyses. Literature, and the imaginative space it traverses, provides the means to interrogate universalities and transcultural intersections in a productive way. The case studies for literary investigation may be fictional, but they are but transfigurations of real cultures, histories, and contexts from which they emerge and are embedded in. As Isabelle Meuret reminds us, '[Literature] is, in the end, the very reflection or imaginary projection of these domains.'[28] Literature can effectively individuate the crisis of mass hunger, contextualize the impact of food insecurity within the social consciousness, and consider the multiple, often contradictory, symbolic meanings of hunger. Food and hunger can be uniquely explored through the rigours of literary analysis by exploring specificity, as well as universality. It is literature's capacity to do both that determines its productive contribution to the field.

Although it appears to be a counterintuitive approach to discussing food and the body – from a conventional standpoint, the very stuff of materiality and the biological – the formation of food, the process of eating, and the symbolic repercussions of hunger can only be grasped if food is theorized as a form of communication, as an economy of signs replete with interrelated representational and symbolic meanings. 'As a language, food-related behaviour is multifaceted, uniting both biological desire for food and the more complicated longing for less tangible "foods" that humans crave: acceptance, respect, love, support, security, self- determination.'[29] The alimentary produces meanings in every stage of production and consumption. 'Cooking is a language through which that society unconsciously reveals its structure.'[30] Mary Douglas states in 'Deciphering a Meal':

> A code affords a general set of possibilities for sending particular messages. If food is treated as a code, the messages it encodes will be found in the pattern of social relations being expressed. The message is about different

degrees of hierarchy, inclusion and exclusion, boundaries and transactions across the boundaries.[31]

Food, like language, relies on a complex system of difference, similar to a semiotic system. In the same way that subjects are interpellated into ontological states of being through language – given meaning and value within the structures of language and power – so is eating an analogous system of subject formation. It is its resemblance to language that facilitates food's potency as a rich symbol for literary analysis. In *Toward a Psychosociology of Contemporary Food Consumption*, Roland Barthes outlines this understanding of food and consumption:

> For what is food? It is not only a collection of products that can be used for statistical or nutritional studies. It is also, and at the same time, a system of communication, a body of images, a protocol of usages, situations, and behavior [...] When he buys an item of food, consumes it, or serves it, modern man does not manipulate a simple object in a purely transitive fashion; this item of food sums up and transmits a situation; it constitutes information; it signifies. That is to say that it is not just an indicator of a set of more or less conscious motivations, but that it is a real sign, perhaps the functional unit of a system of communication.[32]

Food is comparable to a highly complex communication system. Like language, an item of food can be understood as the basic unit of language: a sign. The communicative and linguistic properties of food are vital in understanding the forms of the hunger strike, as well as what is being said through the somatic meanings contained within. This project reads food as a language and interrogates and problematizes the politics of language (in particular, the Anglophone language of the colonizer) as it is critiqued within postcolonial theory. In the imperial and (post)colonial context, food operates as a system that communicates colonial ideology – operating through and upon the body. The language of colonialism encases and produces the colonialized and postcolonial body, and my work examines the ways in which this process is achieved through the symbols and signs of food and eating.

I insert my texts and hunger-striking bodies into historical trajectories that reflect and intersect with accounts of national hunger and the forms of its privations. I trace a broad history of the hungry body by locating it in a past characterized by food insecurity. I consider how this insecurity is in turn affected and/or created by imperial legacies, leading all the way to present forms of global capitalism and the food politics and poetics it produces. These historical narratives of power create the text of the starving body – of both hunger strikers and victims of food insecurity. I explore how historical contexts of hunger may have influenced later representations of hunger texts studied in this work, and why hunger specifically is the mode used to explore matters of power and domination in the colonial/postcolonial moment. Does a history of food insecurity create a collective and lasting memory of hunger that then leaks out in postcolonial writing? Certainly, famine may be read as a historical or collective trauma – and in some instances, this study makes these kinds of connections. However, in most cases, the link between the figure of the hunger striker and a history of hunger is not as straightforwardly framed as a trauma. Trauma Studies is utilized in the final chapter of this work, but it is not a primary theoretical framework used throughout. Instead, the staging of hunger by the individual body is focalized through a reading of material and historical legacies of imperialism, articulated in the inequalities of power, value, and representation. The hunger strike is a response to a fractured subject position that characterizes the postcolonial and colonial experience. This fractured subjectivity traces its origin to an oppressive power differential in the past and/or is a retort to repressive ideologies in the present. As an appropriate field in which to study the nature of narrative – including historical narratives – literary studies examine the ways in which the materiality of the body, the textuality of history, and the historicity of imaginative fiction intersect.

The Body, Food, and Identity

Food can connect us not just within material food systems and economies but socially, culturally, and symbolically. Eating is necessary for survival. It is a universalizing activity and produces and reproduces the metaphysical structures of the self through the body. Through food and eating, the individual is inserted into a densely constructed economy of social and cultural meaning. 'Intensely social, boringly mundane, simple or complicated, at times eating seemingly connects to the very core of ourselves, at others it is just a drudge activity necessary to keep body and soul together.'[33] Food is particularly useful when thinking through the subject because it facilitates an interrogation of the intangibility of the self and the materiality of the body. It brings into scrutiny the body in society. 'The body, a social product which is the only tangible manifestation of the "person," is commonly perceived as the most natural expression of innermost nature.'[34] As a topic, food is fertile ground for analysis due to its ability to affect the conditions of the body – physically and literally, which goes without saying, but also metaphysically and psychologically. This is due to the tight union created, and sometimes taken for granted, between the body and identity. As such, eating is an expression or technology of identity and the body. Food 'structures what counts as a person in our culture,'[35] and the body is the ground for its various significations. In examining seemingly quotidian acts of consumption, one gains access to a medium of inquiry that connects issues of identity formation with community and culture.

Conceptions of the body are historically determined, as well as culturally. Not all bodies in a society are created equal. How bodies are understood is constantly in flux and changes over time. Theorists such as Judith Butler and Michel Foucault have cemented the notion that bodies are socially constructed, and this has been widely embraced in western discourses of the body.[36] They position the body as an object that acquires meaning and definition through interactions with the world around it. The body is interpellated and (self)-policed into being. The material body is always already filtered through the determinations of culture. Through repetitive performance and self-discipline, the body is an emergent locus of the self — but discursively predetermined. 'All bodies are generated from the beginning of their social existence (and there is no existence that is not social), which means that there is no "natural body" that pre-exists its cultural inscription.'[37] These theories dovetail with Pierre Bourdieu's concept of the habitus, which asserts that repetitive socially received behaviours, tastes, and choices constitute identity. Habitus is 'the way society becomes deposited in persons in the form of lasting dispositions, or trained capacities and structured propensities to think, feel and act in determinant ways, which then guide them.'[38] The identities described by Bourdieu are modern bodies — post-Enlightenment and individualized.

Bourdieu states that food choice is not an expression of innate preferences. Food choices, like other aesthetic proclivities, are socially informed. In *Distinction*, Bourdieu demonstrates how social class, education, and background impact upon seemingly automatic and essentialized aesthetic choices, and how this choice in turn reinforces and reproduces social position. Utilizing Marxist frameworks, he examines how the control of capital determines aesthetic taste and how these tastes are utilized to clearly identify, and demarcate, disparate social groups. Bourdieu's seminal work interrogates the link between identity and food: how food choice can be an expression of identity, and how identities are a product of the tradition and patterns of aesthetics they are born into. 'Taste distinguishes in an essential way, since taste is the basis of all that one has — people and things — and all that one is for others, whereby one classifies oneself and is classified by others.'[39] Personal preferences, and unique subjectivities, are expressed by choice, and as culturally constructed choice is literally ingested, taken into the site of identity — the body — it is a potent tool for distinguishing self from other. A subject acquires his/her unique identity through this process of differentiation, based on aesthetic choices that seemingly naturally flow out from the site of subjectivity: the body:

> Social subjects, classified by their classifications, distinguish themselves by the distinctions they make, between the beautiful and the ugly, the distinguished and the vulgar, in which their position in the objective classifications is expressed or betrayed. And statistical analysis does indeed show that oppositions similar in structure to those found in cultural practices also appear in eating habits.[40]

By positioning himself/herself in this endless system characterized by oppositional sets of desirable and undesirable qualities, the subject comes into being. Bodies/identity are considered and constructed as distinctively 'me' or 'you' because 'I' express a seemingly innate identity that prefers French cuisine while 'you' prefer American cooking. The way bodies/subjects are apprehended in Bourdieu's work is repeated by other theorists. For example, Ellmann states this familiar means of subject formation in her book *Hunger Artists*, but through a psychoanalytic framework.[41] 'Eating is the origin of subjectivity. For it is by ingesting the external world that the subject establishes his body as his own, distinguishing its inside form its outside.'[42] This theorizing, again, locates subjectivity in/on the body; eating acts as the process through which bodies are separated and defined against one another, and yet in this process of differentiation, disparate bodies are united in a simultaneous process of identity formation.

Although Bourdieu's *Distinction* is still received as a foundational text in the field of Food Studies, especially concerning how food structures somatic identity, it has not been without its critics. His work is a useful point of entry to understanding how food and eating construct the body and self, and how culture produces and reproduces bodies and tastes. Bourdieu makes sure to indicate that appropriate food choices are socially directed, foregrounding the constructiveness and cultural specificity of his work. *Distinction* is a reading of French aesthetics and cultural practice over a specific time period. Bourdieu acknowledges that: 'The aesthetic disposition demanded by the products of a highly autonomous field of production is inseparable from a certain cultural competence.'[43] So, for example, his assertions in relation to class-specific meal preferences are undermined by the rise of global contemporary restaurant culture. Not only has the restaurant experience become more widely available to a variety of social groups and classes, but the dining-out experience is more than simply an occasional special gastronomic treat; it is a space where cultures, discourses, and values are replicated, transformed, and consumed. Restaurant dining facilitates the removal of socially specific objects – in this instance, food – from their socio-cultural origin, and then repackages and retails them to a new, broad audience. In this context, the meaning of certain foods, (which, according to Bourdieu, is an indication of class, gender, and other identity markers), evolves. Now the preferred *coq au vin* of the farm worker is readily available in restaurants alongside *bouillabaisse*, the favoured dish amongst the French professional class. This change has occurred not just in France but on a global scale, as can be evidenced by the cosmopolitan lure of foreign and 'exotic' cuisine. In his examination of postcolonial Belize food culture, '"Real Belizean Food": Building Local Identity in the Transnational Caribbean,' Richard Wilk notes:

> Taste and preferences are therefore always polysemic in Belize; there is no overwhelming order imposed by a strict hierarchy of capital. Fashion exists not in Bourdieu's two-dimensional space, linked to underlying variation in class, but in a multidimensional space tied to a series of different sources of power inside and outside Belizean society. These other kinds of power include access to foreign culture through relatives, visits, tourism, or temporary migration […] Foreign goods create local identity on a global stage.[44]

This article notes the distinct food habits and practices of Belizean culture, but it is useful as a comparison and critique of Bourdieu's analysis of French culture because it considers the influence of globalized capital – cultural and otherwise.

> Finally, in the modern dietary constitution, what has become of the traditional association between food and regional and national identity? It is a truism to say that our diet has become significantly globalized, and one cultural consequence of that is a hollowing out of the ancient sentiment that custom is a second nature. We eat everything, from everywhere, at any time of year.[45]

This example necessitates an acknowledgement of globalization when thinking through food, a theme explored in Chapter 1 of this book on Kiran Desai's *The Inheritance of Loss*, where food and eating are experienced by characters as a rootlessness and produce uncertain subjectivities, due to their insertion into a delocalized experience of a postcolonial, global food system. These examples also indicate the necessity of reading food as an increasingly internationalized and polysemic language that can be productively framed by the theories and discourses of postcolonial and theories of world literature.

Another critique of Bourdieu's utilization of food as a classificatory system is the rigidity of the categories produced by his theories. In her book *Carnal Appetites*, Probyn takes the established trends of scholarly approaches to food and eating to task. She critiques Bourdieu's work on the social reproduction of aesthetic taste by claiming that his analysis produces a kind of rigidity; that he immobilizes the category of the body, deploying it as a passive container for social meaning. Bourdieu claims that 'the body is the most indisputable materialization of class taste.'[46] But in this classificatory role, it becomes inert, simply a recipient of representation.

> Due to Bourdieu's efforts to avoid the twin perils of objectivism and subjectivism, 'to really get beyond the artificial opposition that is established between structures and representations,' the body emerges as a principle that follows the orders of its past. The body is 'sign-bearing, sign-wearing,' the producer of signs which are physically marked by the relationship to the body, yet it remains strangely inert. The body that eats is in the end eaten by the overdeterminations of culture.[47]

On the other hand, Foucault approaches Bourdieu's schematic of classification differently. He presents the relations between the object of the body and the cultural machinations that determine it as interdependent and fluid, stating: 'the individual is not to be conceived as a sort of elementary nucleus [...] on which power comes to fasten [...] In fact, it is already one of the prime effects of power that certain bodies, certain gestures, certain discourses, certain desires, come to be identified and constituted as individuals.'[48] Here, Foucault unsettles Bourdieu's fixed category of the body; it is revitalized beyond the inertia imposed by Bourdieu. It is a fluid category rather than a static container for meaning and is granted a more active role in not only its own classification but also in the classification of that which is placed in opposition to it. Individuals are not fixed entities, nor are our bodies. If perceptions of the body are constantly shifting, and our behaviours cannot be read as essential or 'natural,' then they can be altered by reconfiguring the forms of discourse through which they are expressed.

> In some ways this is a cognitive approach, one that recognizes that bodies interact with the world around, have bodily functions, have limits to their actions, and that these experiential actions will have impact on how people engage with their own bodies.[49]

In this theoretical imagining, food is not the unilateral ingestion of the social into the body, and contrasts with the modern conception of the western body as atomized; a product of the social and discourse.

> [This] is the problem of placing one object, the body, alongside another, food. This is the problematic structure that rules much commentary on the eating body. As we eat or are eaten by the social, the body is either placed before food, and eating confirms its status, or eating is superimposed upon the body as a separate structure.[50]

There are ways to situate the body between essentialism and constructivism, so the body can and does have more agency. We can theorize a body that is both socially determined and can interrogate the limits of the social.

Foucault echoes Probyn's concerns and presents alternative modes of thinking through food and bodies. In a 1983 interview,[51] he argues that modern liberation movements should return to the ancient model of ethics – of which diet was a primary concern – as aesthetics or transformative practice. Foucault would say the body and what is outside of the body – other objects, other subjects, culture, and the world itself – are situated in complex sets of relations that are always in flux. This emergent model grants the category of the body a greater participatory role in its own formation and accommodates more dynamic figurations than Bourdieu suggests. 'Although we are disciplined in what we eat by our upbringings, media, agribusiness, and by government agencies, we may resist these disciplines through counter-cuisines that are, in fact, a form of political resistance to disciplinary power.'[52] Bourdieu's theory of taste is a vital step towards identifying a central occupation of this book, but I pivot towards Foucault's suggestion that the body might be apprehended as a site of embodied agency. 'In this view, the senses, the physical and the gendered experiential body, participate actively in the formation of personhood and society.'[53] Foucault's imagining of the body provides a helpful point of departure for some of the research questions this work explores, particularly in relation to the politics and logic of the hunger strike, and postcolonial hunger. I read the hunger strike, or hunger as protest, as an attempt to interrogate these multiple and often competing concepts of the body and consider embodied agency as a site of political resistance as well as biopolitical control.

Body Theory has been an explicit interest of disciplines such as anthropology and sociology for some time, but the past 30 years have seen theories of the body become more relevant across the humanities. Historical, sociological, and ethnographic theories of the body have been vital in my understanding of the hunger in this study. A useful framework for understanding somatic ontology is expressed through notions of the 'open' or 'closed' body; iterations of one – the 'closed' body – have already been briefly described here. Closed bodies are the dominant somatic formation in western discourse. This body is bounded, individuated, and contained. In this model, the individual body is the site of individual identity, separate and distinct from other bounded, closed bodies. This body is concerned with what is incorporated within the body/self, and what is ejected or rejected from it – food being a primary object of mediation between the two. This inside/outside dichotomy is what constitutes the body in society. What is ingested is taken permanently inside the body and the borders of the body hold fast to facilitate the transformation of the body, and thus the self. Different theorists mark the historical emergence of this atomized body at different points – some indicate the change occurred in the Middle Ages,[54] others trace it during the Renaissance period,[55] while others insist that it is solidified during the Industrial Revolution.[56] This broad historical range indicates that, in fact, the idea of the closed body emerged in stages alongside other ontologies of the body – the primary competing version being that of the 'open' body.

The open body positions the community within the individual, and its somatic borders are constantly in flux. The individual body operates as synecdoche, representative of the whole community, and the community in turn constructs social identity in the individual body. The open body concept is more commonly associated with non-western communal cultures, although European cultures and societies have evidenced the open body model, both in the past and present. The ambivalent relationship between these two ontologies – the closed and the open – suggests that, in fact:

> it is best to consider this as a spectrum, with the potential to have multiple aspects of these potential beings to fracture the universalization of the modern concept of the bounded individual and seek other worldviews of the self, the body, and the community.[57]

The open body model is one of permeability. In this understanding, identity is predicated on a dualism of self and other, and becoming is a process of negotiating these two poles.[58] These concepts of the body – and how the body and its boundaries may be affected by food or starvation – are important to my understanding of the starving bodies and subjectivities in this study. The hunger striker's protest takes the form of one-of-many – thus, the 'open' body of the striker stands in for a larger community in whose name he/she enacts resistance. He/she enacts the protest synecdochally for an oppressed group. Conversely, sometimes the hunger strike is an effort to shore up the borders of the body, to solidify and police them – an effort to be ontologically 'closed' off from the suppressive context that imprisons it. The hunger striker also relies on the boundaries between the self and the other to give the logic of the strike its structure – in many ways, a hunger strike is always a strike *against* something or someone, so the body in this instance is conceived of as open *and* closed: closed bodies relate in opposition, but the permeability and mutability of the hunger strike rely on creating a spectacle of suffering that incorporates the other into its significatory logic to achieve its political goals – it forces an alignment of affect between striker and audience. Somatic boundaries are necessary for this kind of oppositional stance, but they are not entirely closed off. Through hunger protests, the body emphasizes relations to other bodies and thus can and does affect its own significations, as well as those bodies outside of it. The hunger strike must be read on multiple levels and through various body theories to fully appreciate its various meanings and structures.

Affect theory, exemplified in the works of Patricia Clough, Eve Sedgewick, and Laurent Berlant provides a useful theoretical entry point into the notion of this fluid body:

> [I]ntegral to a body's perceptual becoming (always becoming otherwise, however subtly, than what it already is), pulled beyond its seeming surface-boundedness by way of its relation to, indeed its composition through, the forces of encounter. With affect, a body is as much outside itself as in itself – webbed in its relations – until ultimately such firm distinctions cease to matter.[59]

Affect theory facilitates the exploration of a multifaceted body. It considers the agency of the affective material body, but also the body-as-concept within discourse. Thus, the body, communal and individual, is:

> constituted, not outside but within representation [...] not as a second-order mirror held up to what already exists, but as that form of representation which is able to constitute us as new kinds of subjects, and thereby enable us to discover places from which to speak.[60]

A variety of theories of the body are engaged in this work in order to understand the complexities of a hunger strike, and how it can be read as an intervention in the politics of identity. Affect theory provides a natural theoretical home for Food Studies disciplines because it connects the sensory phenomenology of food and eating – taste, smell, touch, sight, and even sound – with the affective body. Memory, nostalgia, emotion is stirred up in the body in affective as well as cognitive ways through food. In the alimentary, 'senses and affect bleed into one another.'[61] Through this reading, the body is imagined as a 'nexus of finely interlaced force fields [...] that register links between perception, affect, the senses, and emotions.'[62] The affective process of starving-as-spectacle co-opts the spectator in a mutual register of pain and suffering, thus influencing not only the body-as-self but also the unstable body-self as constructed between self and other.

This book considers these questions of the body and self through the medium of postcolonial writing, and these questions come to bear specifically on the colonial/postcolonial subject. The hunger strike is conceived as a somatic practice or technique. It is a response meant to transform the postcolonial subject's experience and self. The success and/or failure of this practice is considered in each chapter. My readings demonstrate that hunger strategies are characterized by ambivalence. They can be read as oppressive, self-inflicted forms of biopolitical control, but there are still moments of political value contained within these embodied practices. The usefulness of these instances of hunger and starvation are generally not gleaned by the subject themselves, but rather by third parties who have access to and read the hungry body-as-text in politically productive ways. The violence of starvation, the material and somatic consequences of this self-styled protest, is extreme enough that although the hunger striker may have been staging a protest in response to a social and ideological injustice, they do not reap the benefits of their own suffering, subject as they are to the mortality of the starving body. However, their physical suffering – and even death – can produce meaning after the body is gone, for the communities in whose name they protest or the societies they leave behind.

Body/Mind Dualism

This book reads the body as a text, one that can be read in a variety of ways. I consider how the body is constructed by contextually specific discourses – and how these discourses affect the way in which the material body is conceived. This is not to deny the materiality of the body or to argue that the body is merely a textual object. As stated above, the materiality of the body can sometimes be obscured by the overdeterminations of culture, which can prevent a fully conceived notion of the embodied self. My readings of hunger-striking bodies acknowledge the irrefutable and corporeal realities of the human body. A starving body is not an inert object or an empty site of signification, but rather a violently active site of both meaning and pain. This project seeks to understand how discourses shape the meaning and value of the fleshy body, but also how the body might resist the power of discourse. The postcolonial subject is constructed at the intersecting site of several somatic and ontological paradigms, constructed by colonial discourse, and reinforced by its literature and writing. The postcolonial/colonial body can be conceived of as a text that is created by, and inserted into, colonial and postcolonial discourses. My work examines how hunger strikers attempt to rewrite the colonial body through imposed hunger, considers the usefulness of such an activity, and assesses the origins and forms of this protest. It seeks to understand how categories of the body and mind, of nature and culture, are deployed within various postcolonial texts, and how these categories are brought to bear on, and negotiated by, the colonial/postcolonial subject.

The body/mind binary is an essential framework for this study's analytic approach. This duality underpins much of western philosophical thought, traced back to Plato's assertions that the haptic and sensual pleasures of the body were corrupt compared with the loftier workings of the mind, and the pursuit of knowledge. He theorized that the body could be deceived, led by its baser desires, and only the soul was capable of recognizing truth and the purity of the world. However, the body/mind divide is more commonly associated with seventeenth-century philosopher René Descartes, who theorized a stark division between body and mind. The rise and acceptance of Cartesian dualism coincide with the European Renaissance period's atomization of society:

> The mind/body divide's popularity was advanced in the Renaissance concept of civilizing, which promoted a process of individualization, seen both in society's breaking up into specializations and also in the concept of the individual being disarticulated from other individuals around them.[63]

The body/mind binary, therefore, was integral to the emergence of the 'closed' bounded body and arose alongside social formations that encouraged individual somatic practices, that could and should be practised alone – private and public became more clearly demarcated. An alimentary example of this was the greater emphasis placed on mealtimes and comportment at the dinner table. Instead of communal meals that involved sharing food from the same plates, eating with hands, and a commensality characterized by a breaking down of boundaries between the self and other, the table and its mores became strictly monitored and increasingly prescriptive. The dinner table became a site of self-control, and individuals were responsible for their own behaviours – distinguishing proper from improper. Individuals were tasked with 'civilizing' their bodies in the alimentary space. These individuating ideals were built upon during the Enlightenment, and found easy expression in the logic of the European colonial 'civilizing' mission:

> Enlightenment thinkers built on this Cartesian dualism of body and mind, extending it to radically separate thinking 'man' from unthinking 'nature.' Distinguishing European man's presumed capacity for mental reason from the supposedly animalistic nature-bound bodily impulses of women and the colonized, Enlightenment thinkers created a powerful justification for colonial expansion and the exploitation of labor and environment. These liberal political philosophies continue to underpin imperial control of land and labor within the international system of nation-states.[64]

In these western-centric discourses, the mind is the privileged site in the body/mind binary, as is the individual over the communal body. While Europeans were civilizing their own bodies along Cartesian lines, this same civilizing impulse was exported to their colonies. The colonizing force is constructed as progressive – emblematic of the logic of the mind. The native is received as an irrational site of bodily sensation – a creature of the somatic. The colonial mission rationalized its dominion over colonial spaces and bodies by deploying this civilizing discourse, mobilizing the concept of the closed, individuated body and the hierarchical logic of rational Enlightenment thought. Inserted into the body/mind formula, the native/colonial body is posited as natural and immutable, whereas the mind — and the associated 'logical' colonial civilizing force — is seen as distinct from this — a social product and changeable. Thus, native bodies are doomed to remain static and savage, controlled by somatic urges and characterized by a lack of rational intelligence. The colonizer's body, on the other hand, is controlled and controllable — it can alter itself with rational will, and continue to improve, in line with the progressive ideals of Modernity. In this structure, the colonial body is erased and rewritten as a symbol of animalism and savagery. Complex histories and identities are deleted in favour of a flattened teleology of an eternally savage land and people. The colonial/postcolonial body becomes an empty signifier, the location of othering, and the target of the colonial civilizing project. Yet the native body is also the carrier of a range of specific characterizations, necessary in constructing and defining the boundary between the metropolitan centre and the orientalized geography of the colony.

In the case of the colonial/postcolonial body, the body (or 'nature' in the corresponding nature/culture divide) is seen as the sticky/stuck category that cannot be changed. It is an innate body, yet still subject to revisions by the civilizing discourses of colonialism. This presents a contradiction, considering that the body/mind binary relies on the native body remaining constant and unchangeable as an uncivilized site of emotion and sensuousness, so that the mind (and thus colonizer) can occupy the privileged position of mutability and Modernity. These formulations and contradictions persist in the racist ideologies of the neocolonial present. The 'native' colonized body is a densely constructed political site, produced by colonial and postcolonial encounters. My project examines how it is necessarily sustained in and for the project of colonization, and the ways in which the site of the body is controlled by colonial ideologies to perpetuate the binaries of imperialism. Hunger strikers attempt to take back control by asserting agency over their bodies — even if that agency takes the form of starvation, and that starvation is structured using racist metropole discourse that privileges mind over body. My project considers and imagines how colonial discourses are internalized within, and structure, nationalist and anti-colonial narratives, and concurrently the postcolonial body itself and its hunger protests. It is the contradictions that arise from these competing discourses that place the colonial/postcolonial body in crisis.

Theories of Race and the Colonial/Postcolonial Body

Formulated by Scottish thinkers such as Adam Smith and William Robertson, stadial theory was popularized in the eighteenth century; according to Roxanne Wheeler, it 'provided the most important rubric for thinking about human differences in the eighteenth century.'[65] It posited that an individual's distinctive traits and behaviours were due to their position along a progressive historical trajectory that was configured as a teleological continuum from primitivism to civilization. These theories 'understood national or racial characteristics to be contingent and malleable.'[66] This racial discourse dovetailed with climatic and humoral theories, the former positing that human characteristics – both physiological and psychological – were a result of climatic influences such as the weather and soil quality, the latter inherited from classical understandings of the body as controlled and defined by humoral theory. These theories of the body and difference suggest that malleability was important particularly in relation to bodily identity in the eighteenth century, demonstrating that subjectivity then was more elastic than it is today in terms of how we conceive of our bodies. 'The several examples of the ease with which visible change occurred to individual bodies and to the body politic suggest that Britons' understanding of complexion, the body, and identity was far more fluid than ours today.'[67] The extent to which bodies were able to change was dependent on a number of factors both within and outside the bodies' influence, but there remained a sense that bodies – and so identities as attached to these bodies – could change and did so.

By the mid-nineteenth century, biological theories of race had become a key concept in metropolitan thinking about individual and group identity, moving on from stadial conceptualizations of bodily difference and alterability: 'those who focused on the notion of permanent physical differences which were inherited and which distinguished groups or races of people one from another.'[68] Biological theories of race shifted to a more fixed idea of indexed difference, and in the context of the colonial encounter, these differences were drawn along racial lines. Bodies were classified according to the racial categories that we have roughly inherited and deploy today – unlike previous classificatory methods that included categories such as clothing and religious affiliation. Biological determinism regarded different races as organically distinct and unable to successfully integrate. However, the more fluid stadial framework continued to influence understandings of identity and the body, operating concurrently with more rigid classificatory practices, producing a contradictory set of taxonomies that were deployed as and when required. 'Biological racism and cultural differentialism, therefore, constitute not two different systems, but racism's two registers,'[69] and these two discourses operated simultaneously, with 'the cultural slipping into the biological and vice versa.'[70] The imperial project was supported by both of these theories of race, and both were ambivalently deployed to serve a number of justificatory needs. Theories of teleological development encouraged the 'benevolent' discourses of 'the white man's burden' – a civilizing imperialism that was charged by moral imperative. Simultaneously, the continued violence and exploitation that characterized imperial relations were bolstered by narratives of the biologically inferior subaltern. In these instances, the continual process of discipline and control that determined colonial rule could be seen as a kind of natural order, with superior species sitting on the apex of a hierarchical racial pyramid – human – and the races increasingly approaching animals further down the hierarchy.

This hierarchal system is directly traceable to the colonial encounter, focalized through an alimentary logic:

> The right foods – those to which the colonists were accustomed, notably wheat bread and wine – would, it was thought, protect the colonial body from the physiological risks of the New World environment, while eating local foodstuffs would transform it into the flawed native body. Native foods were responsible for native humours and temperaments. You are indeed what you eat, and the colonial enterprise was understood to depend upon maintaining the constitutional difference between colonists and colonized.[71]

Over the course of imperialism, however, these fears of contamination were neutralized by an Orientalized economy of desire and taste. Exotic objects of consumption became fashionable and circulated widely within the Empire's economy. Products like spices, tea, coffee, and sugar became staples of the English diet. These newly configured foodways reserved for the colonizer the luxury of a diverse diet, sanitized through the process of border crossing into the metropole, but also supported the imperial mission of civilizing through colonialism – further justification to secure these imported items. Colonialism reconfigured the 'fantasmatic landscapes and the sensorium of colonizer and colonized, generating new experiences of desire, taste, disgust, and appetite and new technologies of the embodied self.'[72] Stadial models of race were still a mainstay of colonial discourses, even within the context of consumption. Take, for example, Isabella Beeton's take on global dining practices:

> Man, it has been said, is a dining animal. Creatures of the inferior races eat and drink; man only dines [...] It is equally true that some races of men do not dine any more than the tiger or the vulture. It is not a dinner at which sits the aboriginal Australian, who gnaws his bone half bare and then flings it behind to his squaw. And the Native of Terra-del-Fuego does not dine when he gets his morsel of red clay. Dining is the privilege of civilization. The rank which a people occupy in the grand scale may be measured by their way of taking their meals, as well as by their way of treating their women. The nation which knows how to dine has learnt the leading lesson of progress.[73]

It is clear from this excerpt of *How to Dine, Dinners, and Dining* that Beeton reads dining practices through a stadial model of civilization. Fine dining is viewed as a marker of civilization, while primitivism is associated with animalistic modes of consumption. Even the mistress of the house is responsible for the reproduction of colonial racial discourse within the domestic setting. 'Cookery books tracked contemporary thinking about racial identity and how it was configured in their ongoing exploration of what defined the nation and who could be part of it.'[74] Beeton is less clear as to whether it is possible to move along the spectrum of civilization she describes. The inclusion of the savage dining practices of other cultures serves as a cautionary tale, persuading metropole and colonial readers to aspire towards civilization even within the domestic particularities of household management. The notably animal-like qualities of those she denigrates suggest that this passage should also be read through the inflections of biological determinism, implying that other races were somehow so essentially different to the British that they may as well be a different species. Thus 'racism's two registers' bolster one another while operating ambivalently in colonial discourses of eating and food. Beeton's household manuals were key texts in the dissemination of normative racial thinking, and often contained recipes of foreign cuisines which were a result of increased colonial encounters and cultural cross-fertilization. In most cases, the recipes of 'the inferior races' were consumed unproblematically by the metropole, shored up by the racial discourses of the time: stadial theories placed the English as a more civilized station along the continuum of racial development. They could not de-evolve from that station to a lesser position through consumption, as biological theories placed them securely in a category of racial superiority that could not be contaminated by the food of lesser races. Whiteness, as a category, was able to negotiate the contradictions posed by these competing racial theories without difficulty.

Humoral theories – most closely associated with Hippocratic rhetoric, although increasingly rendered obsolete by nineteenth-century racial and bodily classificatory systems – still found expression in the way the body was constructed during the colonial period. 'Though introduced in Antiquity, this model exerted its influence up until the nineteenth century in scholarly dietetics (and beyond in quotidian life).'[75] According to humoral theory, food had specific transformative attributes. Ingesting these foods meant assimilating these attributes into the body, influencing relations of the four humours – black bile, yellow bile, blood, and phlegm – that made up the body, and in turn their relationship with the organs. These theories proved elastic, and they were able to accommodate some elements of the new, scientific knowledge on nutrition that emerged in the nineteenth century. Thus, Brillat-Savarin's oft-paraphrased aphorism, 'Tell me what kind of food you eat, and I will tell you what kind of man you are,'[76] solidified during the nineteenth century. The division between the self and the body deepened during this period, and food and eating became technologies of the body. 'Society expected people to be rational and controlled, and to possess the detailed knowledge and the strength of will to eat according to the guidelines of rational nutrition.'[77] This rationality comfortably intersected with stadial notions of civilization, and the cross-cultural fertilizations that occurred as a result of colonial encounters – particularly around food commodities – were inserted into this framework. Colonial markets and the introduction of new consumer products into English households and businesses reified this function of identity as located in the body and subject to controlled transformations. According to Edward Said, European rationality asserted that 'the typical materiality of an object could be transformed from mere spectacle to a precise measurement of characteristic elements.'[78] Several historical commodity studies – like Sidney Mintz's *Sweetness and Power* – demonstrate the clear links between global markets and the formation of the self through alimentary practices. 'Tobacco, sugar and tea were the first objects within capitalism that conveyed with their use the complex idea that one could become different by consuming differently.'[79] Access to this transformative agency, however, was limited to the metropole and its citizens, and these theories seemed to operate contextually and with particular attention to the body being considered – some bodies, it seemed, could never change, even though colonized bodies 'were more often than not incarnated in bodies whose appetites, expressions and comings and goings had to be rigorously fashioned.'[80] Stadial theories, humoral theories, and then later – and simultaneously – rational and biopolitical figurations of the body worked in specific ways depending on the body in question. The white, male, and rational body emerged as a somatic category that was able to manoeuvre and exist within the paradoxical matrix of these racial, bodily, and alimentary theories, while retaining a position of power and privilege. Other bodies – colonial or subaltern bodies – had the difficult task of occupying a contradictory space in these racial formulations.

Classificatory processes focused on the body, and on race, as the site at which the self was defined:

> It was colonial encounters which produced a new category, race, the meaning of which, like those of class and gender, have always been contested and challenged. The Enlightenment inaugurated a debate about racial types and natural scientists began to make a new object of study, that is, the human race. They laboured to produce a schema out of the immense varieties of human life.[81]

The bodies of colonial subjects – or 'native' bodies – were, at this time, the focus of much consideration in the form of empirical inquiry. These Othered bodies became the material site for a dialectic that defined their position under England's imperial rule, and then later in the postcolonial period. The increasingly visible and scrutinized native body was subject to both stadial and biological discourses of race, with one approach constantly slipping into another. However, unlike white bodies, the colonial body occupied a paradoxical position in the intersection of these different racial theories. Discourses of development reinforced stadial theories of progress and betterment, providing a rationale for civilizing colonial rule and education, and for encouraging natives to better themselves. Simultaneously, biological classifications meant that colonial subjects could never truly 'improve' and move along the continuum of civilization. They would always, to some extent, remain savage, remain animal. The 'Noble Savage' trope defined the limits of the native's access to knowledge. The colonial and even postcolonial body is positioned within these contradictory racial theories. This accounts for the internalized racism that Frantz Fanon describes as a marker of the colonial subject. The colonial subject was both indoctrinated to accept colonial rule as the natural order and taught to aspire to become like the colonial master, to approximate whiteness as much as possible. Cuisine was one area in which colonized subjects could attempt this, but their engagement with colonial alimentary practices was unable to fully elevate them into the same racial or social category of those in power, however much they tried, due to their essential biological backwardness. The colonial subject was ideologically conditioned to hate the inferiority represented by the native body, and attempted to transcend it, but the goalpost for achieving this ideal was structurally impossible to reach due to biologically determined concepts of racial identity. This internal contradiction is represented in the self-loathing or double consciousnesses that Fanon explores in his work, the internal, paradoxical psychodynamic proving to be an irresolvable subject position, and ultimately beneficial to the ruling colonial masters.

These theories of race and the internalization of colonial ideology explain the contradictions of the colonial hunger strike. The hunger strike is an effort to respond to, and to resolve, the paradoxical condition of the postcolonial body. The hunger strike is, correspondingly, a paradoxical solution to a paradoxical problem. The hunger striker commits self-abnegation as a means of attaining freedom, denying herself/himself in order to deny the colonial master, and essentially destroying the self as a means of finally achieving the humanity denied to native populations under colonial rule. The contradictions of race and body will be explored in greater depth through the hunger strikers examined in this book. The irresolvable contradictions of the native body are tackled by the equally problematic and paradoxical intervention of self-starvation.

Hunger Strikes and Historical Hunger

I read hunger strikes as a response to imprisonment or control. This control can take literal forms but can also be explored through ideas of ideological, social, political, and psychological control. 'The violence of colonialism has always been expressed in the intimate, material terms of controlling bodies, ecologies, and social relations that constitute everyday life.'[82] The bio-politics of control of the postcolonial or native body is, as the Foucauldian terminology reminds us and Fanon reiterates, internalized and self-imposed. The ontology of colonial and postcolonial identity – of double-consciousness – results in self-hatred:

> Every colonized people – in other words, every people in whose soul an inferiority complex has been created by the death and burial of its local cultural originality – finds itself face to face with the language of the civilizing nation; that is, with the culture of the mother country. The colonized is elevated above his jungle status in proportion to his adoption of the mother country's cultural standards.[83]

The resolution to the internalized racism Fanon describes can be traced to the hunger-striking mechanism. The hunger striker attempts to move away from self-identifying with the savagery of the body and instead attempts to transform into the rational agent of the mind. The mind is allied to discourse, language, reason, and control. Within this semiotic framework, a literary analysis can productively intervene into the somatic practice of self-starvation because the hunger striker's tools for reshaping the self are narrative and discursive, alongside the somatic techniques of self-abnegation.

In many ways, the hunger strike is a curious form of dissent, as it is a response to biopolitical control with yet another form of biopolitical control. It is an attempt to escape the identificatory abjection of the somatic side of the body/mind divide that the racialized colonial/postcolonial body is associated with. Animalism, sensuality, and uncontrollable urges characterize the colonial body. The hunger strike is an attempt to 'civilize' the abject corporeal Other/self in the Cartesian formulation, to become the self-controlled rational body, associated with the privileged category of mind. However, the abject Other in this instance is the colonial/postcolonial body itself. The hunger strike's formation is flawed because it structures and locates its resistance within these paradoxical discourses of Enlightenment. These are the same structures that ensnare the colonial/postcolonial body in the damaging and irresolvable racialized binary represented in the Cartesian dualism. Ultimately, the body is inscribed and indivisible from the mind in this ontology. A deconstruction may be possible, but a negation or escape of the colonial/postcolonial self is not possible by using the language of the colonial masters themselves. Thus, the hunger-striking body's invariable conclusion is often self-obliteration.

However, the hunger strike can also contain productive moments of dissent. It can be read as an attempt to reassert the somatic as an interrogative tool as protest. This is a reclamation and re-inscription of the body. Hunger strikes can voice erased colonial histories using their body as text, and by 'making flesh' the violence of these elided narratives. As Michelle Pfeifer states in her analysis of German refugee hunger strikes: 'I use the term becoming flesh to signify how the racialized and abjected body opens an arena for reclamation in which the hunger strikers could articulate a politics of refusal.'[84] The body becomes a site of expression and dissent – the body 'speaks' using a radical lexicon that attempts to overcome the rigid discursive structures of colonial logic and history. This project investigates how historical famines created or accelerated by colonial intervention are recuperated and redeployed in the somatic register of the hunger striker's body. In the self-styled violence undertaken willingly by the native and postcolonial body, the subject rejects the politics of natural or climactic causation generally applied to understandings of widespread hunger and starvation, and asserts a new story. In the radical rejection of the image of the wretched of the earth, and the reclamation of the body-in-pain in a new socio-historical context, the text of the body subverts the politics of empathy produced by popular histories of the Global South that position it as an unlucky or unfortunate geopolitical space in terms of food growth; burdened by uncontrollable, impoverished subjects incapable of appropriate population control. Instead, the starving body becomes an insistent spectacle, to be read in different terms. Although the body produces multiple readings, and cannot be singled to just one, my project is interested in those readings that might afford the colonial body some power and expression, as well as those that do not, and can be read as failure. The aesthetics of pain produced by the hunger strike can give literal shape to the erased histories of imperial violence. This hypervisibility is at once an act of destruction, as well as proclamation of resistance.

The relationship between hunger strikers and language/words is complex, but important for the purposes of this work – particularly as a literary study. Although the hunger strike is a somatic technique, it invariably relies on discursive logic to give it meaning – even a refusal to define a hunger strike is a type of speech. Although in the formulation of the hunger strike it is the body that 'speaks,' hunger strikers will often either substitute food with an overabundance of words/speech, or a lack of it. This may denote two forms of a similar sort of resistance – a vocal hunger striker tries to rewrite the body and its history in more obvious ways (for example, Bobby Sands), while a silent hunger striker allows the body to do the talking. In any event, my account of hunger strikers emphasizes Barthes' insistence that when it comes to texts, the Author really Is Dead: in most cases, the hunger striker remains an inscrutable text – not speaking, speaking indecipherably or silent in death – while those who view/read the strike may find themselves transformed. It may be that only the audience of the hunger strike can truly benefit from it – it can be read productively, but it cannot speak productively. Given the horrors and violence of colonialism, both physical and psychological, this self-obliteration may serve as the only means of voicing the unspeakable. Rather than fixating too much on the speech forms the hunger strikers intend to deploy in their protests ('Can the Subaltern Speak?'), I think it may be more useful to ask the question, 'Can the subaltern be read? And what readings can be gleaned from these starving bodies – productively or otherwise?' Answers to these questions can be mined from the many contradictions and ambiguities contained within the hunger-striking body. Achille Mbembe argues that under conditions of the necropolitical subjugation of life to the power of death, 'the lines between resistance and suicide, sacrifice and redemption, martyrdom and freedom are blurred.'[85] Under the terror of the colonial state, the potential meanings of the hunger strike oscillate uneasily between these boundaries. The hunger strikers in this work evidence these ambivalences through their self-starvation, and although their individual fates end in violence and/or a slow death, their legacy may and does produce alternative scripts for being and knowing for those left behind.

The hunger-striking colonial body is inscribed with a history of hunger and lack. It becomes a canvas on which the imbrications of imperial power can be displayed. With their emaciated bodies, the hunger striker displays and re-performs a history of colonial famine and food insecurity. Throughout this work, each hunger striker is situated within a particular history of food deprivation, and these historical instances are investigated as an outcome of colonial administration and interventions – not, as is popularly received knowledge, because of 'natural' causes, such as climactic or geographic bad luck, or the failure of the colonial state to properly adopt the 'natural' mechanisms of the laissez-faire economy:

Nowhere was this inversion of natural law, with its equation of hunger with moral strength, or the brutal exposure of the violence and inhumanity of colonial rule, more apparent than in the adoption of hunger as a vehicle of political protest through the hunger march and the hunger strike. Both were tactics that appeared to defy nature and the state in equal measure.[86]

Famines are read as national traumas that are articulated through the disfiguration of the colonialized body. A hunger-striking body contains and reflects the intersections of power, food, and history as a radical enunciation of the unspeakable and unspoken horrors of colonialism. Neocolonial global markets structure national economies unequally – rendering food security a common occurrence in developing nations, while citizens in the Global North consume comfortably and with food to spare. In this work, these instances of having and not-having are read through a historical lens that ends in the contemporary moment. The hunger-striking body asserts itself as a means of communicating these complex material histories and asserts a politics of violent agency over the somatic self. Instead of evoking the spectacle of empathy usually associated with the starving subaltern figure, these hungry bodies move from passivity to a self-styled embodied agency.

The evocation of collective memories of food deprivation through the hunger striker's body also invokes productive uses of the open body model, whereby the communal suffering of colonial populations could be writ large on the individual. This strategy transforms the hunger strike into novel forms of anti-colonial speech acts, and as the open body was generally considered more 'primitive' than the closed, individuated body of modern Europe, the hunger strike can also be read as a recuperation of ontological forms that resisted the individuating forces of rational Modernity. These civilizing discourses encouraged somatic surveillance and forms of biopower that were harmful to subaltern figures – subject to Cartesian dualities whose divisions and inequalities they were unable to overcome. Through situating resistance to colonizing discourses through self-violence, the hungry colonial body disrupts the atomizing forces of Enlightenment thinking, turning the boundary-policing biotechnologies of the self-controlled and self-controlling body outward as a spectacle of suffering, insistent on a communal representational politics. The individual starving body contains a multitude of voices and histories, and while the hungry subaltern body can certainly be accused of flattening and performing an Orientalized image of the wretched poor, the mechanism of the protest brings elements of agency and self-representation into stark relief.

Chapter Breakdown

In Chapter 1, I analyze Kiran Desai's 2003 Booker Prize–winning novel *The Inheritance of Loss*. This chapter establishes the role of food and eating in the process of interpellating subjects, as is demonstrated through the various protagonists' relationships with types of food, differing eating practices, and each other. It explores how personal tastes, whose meanings and preferences may at first glance appear essential, are influenced and dictated by a larger hierarchical system of social capital that is constructed as a result of both contemporary socio-economic realities and a historical past defined by colonial domination and food insecurity. Through these connections, the value of food is realized as a distilled result of present social forms and a legacy of colonially fuelled inequality. This chapter also introduces the first hunger-striking body to be analyzed in this work. Nimi's food refusal is characterized by a rejection of both food and words, and, as such, through Nimi I establish how the connections between language and food are forged and disrupted.

The second chapter focuses on J.M. Coetzee's 1983 novel *Life and Times of Michael K*. This chapter explores the relationship between material and representative categories as they are articulated through the novel's plot and the protagonist's relationship with food and hunger. This chapter is concerned not only with the themes and images of food and hunger within the novel's plot, but also examines how these instances comment upon the extra-textual process of reading the novel itself, and whether self-reflective reading practices can be deployed to politicize the novel, and the hunger strike represented within. The hunger strike in this chapter is exemplified by the main protagonist: Michael K. Like Nimi from the previous chapter, he rejects both food and words.

My third chapter examines Tsitsi Dangarembga's novel *Nervous Conditions*. This chapter locates Dangarembga's novel in a specific historical narrative of famine and hunger in Zimbabwe's past. I read the novel's impoverished setting as a direct result of this legacy of food deprivation. In order to tease out this historical connection, I examine the use of food and eating within the narrative present of the novel and demonstrate how the status of food in the text's setting (colonial Rhodesia) is a result of historical colonial domination that contributed towards Rhodesia's problems with national hunger. Gender is considered when discussing the hunger striker in this chapter: a teenage Shona girl named Nyasha. Her hunger is understood as a response to power inequalities as they are expressed through the intersections of colonial domination and patriarchal subjugation. Unlike the previous two hunger strikers, Nyasha rejects food and substitutes it with words. I examine the implications of such a substitution and situate it within the relationship between Anglophone language and knowledge and the historicized colonial body.

The fourth and final chapter differentiates itself from the rest due to its non-fictional nature. Irish Republican Army (IRA) member Bobby Sands underwent a hunger strike in 1981 that eventually claimed his life in Long Kesh prison, located just outside Belfast, after 66 days without food. The strike caused a media frenzy at the time, and Sands remains a prominent figure in the complicated and often tragic socio-political history of sectarian conflict in Northern Ireland. I examine Sands' writings from his time spent in prison. A prolific writer, Sands wrote songs, poetry, essays, and short stories, and kept journals. He kept a special journal during the first 21 days of his hunger strike. Some of his prison writing is fictional – short stories and poems, for example – whereas many of his pieces are clearly meant to be non-fictional accounts of life in prison or about his own life. I examine a selection of all these works in this chapter. The real-life context of this hunger strike is a useful means for considering the operations of the representational in relation to history or life-writing, which is considered less creative and more 'real.' I examine how political and nationalist ideologies that formed in response to the Great Irish Famine can be traced down through history and continue to impact contemporary Northern Irish geopolitics – more specifically, the context of the Troubles. I also provide a close examination of the forms and mechanisms of Sands' hunger strike as they appear in his own writing, and in other critical sources.

Notes

1 James Vernon, *Hunger: A Modern History* (Cambridge, MA: Harvard University Press, 2007), p. 3.
2 Olivia C. Harrison, 'For a Transcolonial Reading of the Contemporary Algerian Novel', *Contemporary French and Francophone Studies*, 20.1 (2016), 102–110 (p. 102).
3 Françoise Lionnet and Shi Shumei. *Minor Transnationalism* (Durham, NC: Duke University Press, 2005).
4 Deepika Bahri, 'Postcolonial Hungers', in *Food and Literature*, ed. by Gitanjali G. Shahani, Cambridge Critical Concepts (Cambridge: Cambridge University Press, 2018), 335–352 (p. 337).
5 Parama Roy, 'Postcolonial Tastes', in *The Cambridge Companion to Literature and Food*, ed. by Michelle J. Coghlan, Cambridge Companions to Literature (Cambridge: Cambridge University Press, 2020), 161–181 (p. 174).
6 Ania Loomba, Colonialism/Postcolonialism: Second Edition (Abingdon: Routledge, 2005), p. 16.
7 See *The Post-Colonial Studies Reader: Second Edition*, ed. Bill Ashcroft and others (London: Routledge, 2006); Leela Gandhi, *Postcolonial Theory: A Critical Introduction* (Edinburgh: Edinburgh University Press, 1998); Ania Loomba, Colonialism/Postcolonialism: Second Edition (Abingdon: Routledge, 2005); Gina Wisker, *Key Concepts in Postcolonial Literature* (Basingstoke: Palgrave Macmillan, 2007); B.J. Moore-Gilbert, *Postcolonial Theory: Contexts, Practices, Politics* (London: Verso, 1997); Bill Ashcroft, *Post-Colonial Transformation* (London: Routledge, 2001); and Neil Lazarus, *The Postcolonial Unconscious* (Cambridge: Cambridge University Press, 2001).
8 Marlon Simmons and George J. Sefa Dei, 'Reframing Anti-Colonial Theory for the Diasporic Context', *Postcolonial Directions in Education*, 1.1 (2012), 67–99 (p. 70).
9 J. Daniel Elam, 'Anticolonialism', *Global South Studies*, 2017, https://globalsouthstudies.as.virginia.edu/key-concepts/anticolonialism [accessed 6 September 2020].
10 Maud Ellmann, *The Hunger Artists: Starving, Writing and Imprisonment* (London: Virago Press, 1993).
11 Susan Bordo, *Unbearable Weight: Feminism, Western Culture and the Body* (Berkeley: University of California Press, 1993).
12 Susan Bordo, *The Flight to Objectivity: Essays on Cartesianism and Culture* (Albany, NY: SUNY Press, 1987).
13 Carole Counihan and Penny Van Esterik. *Food and Culture: A Reader*, 4th ed. (New York: Routledge, 2018).
14 Other readers containing sociological and anthropological essays in the discipline of Food Studies have been published over the same period. For references, see footnotes 9–14.
15 See *Food and Cultural Studies*, ed. by Bob Ashley and others (London: Routledge, 2004); *The Cultural Politics of Food and Eating: A Reader*, ed. by James L. Watson and Melissa L. Caldwell (Malden: Blackwell, 2000); *Routledge International Handbook of*

Food Studies, ed. by Ken Albala (London: Routledge, 1996); Massimo Montanari, *Food Is Culture* (New York: Columbia University Press, 2005); *Food, Health and Identity*, ed. by Pat Caplan (Oxon and New York: Routledge, 1997); *Eating Asian America: A Food Studies Reader*, ed. by Robert Ji-Song Ku and others (New York: New York University Press, 2013); *Eating Culture: The Poetics and Politics of Food*, ed. by Tobias Döring, Markus Heide, and Susanne Mühlheisen (Heidelberg: C. Winter, 2003); and *Kitchen Culture in America: Popular Representations of Food, Gender, and Race*, ed. by Sherrie A. Inness (Philadelphia: University of Pennsylvania Press, 2001).

16 See *Food: The History of Taste*, ed. by Paul Freeman (Berkeley: University of California Press, 2007); Maguelonne Toussaint-Samat, *The History of Food* (West Sussex: Blackwell, 2009); and Rachel Duffet, *The Stomach for Fighting: Food and the Soldiers of the Great War* (Manchester: Manchester University Press, 2012).

17 See *Reel Food: Essays on Food and Film*, ed. by Anne L. Bower (London and New York: Routledge, 2004); and James R. Keller, *Food, Film and Culture: A Genre Study* (California: McFarland and Company, 2006).

18 See *Food and Philosophy: Eat, Think and Be Merry*, ed. by Fritz Allhoff and David Monroe (Oxford and Victoria: Blackwell, 2007); Carolyn Korsmeyer, *Making Sense of Taste: Food and Philosophy* (New York: Cornell University Press, 1999); and Elizabeth Telfer, *Food for Thought: Philosophy and Food* (London: Routledge, 1996).

19 See Giséle Yasmeen, *Bangkok's Foodscape* (Bangkok: White Lotus Press, 2006); and Judith A. Carney, *Black Rice: The African Origins of Rice Cultivation in the Americas* (Cambridge, MA: Harvard University Press, 2001).

20 See *Food and Architecture*, ed. by Karen A. Franck (Chichester: John Wiley & Sons, 2003).

21 Christine A. Hastorf, *The Social Archaeology of Food: Thinking about Eating from Prehistory to the Present* (New York: Cambridge University Press, 2017).

22 Carole Counihan and Penny Van Esterik, 'Why Food? Why Culture? Why Now? Introduction to the Third Edition', in *Food and Culture: A Reader*, 3rd Edition, ed. by Carole Counihan and Penny Van Estrik (New York: Routledge, 2012), 1–15 (p. 2).

23 Note that Food Studies concerns itself with the social and cultural meanings of food and related topics, and issues such as nutrition, agriculture, gastronomy, and the culinary arts do not generally fall under the remit of Food Studies, unless they are being critically discussed in regard to their cultural impact, meanings, and ramifications.

24 Elspeth Probyn, *Carnal Appetites: FoodSexIdentities* (London: Routledge, 2000), p. 26.

25 See Charlotte Boyce and Joan Fitzpatrick, *A History of Food and Literature* (London: Routledge, 2017); *Cooking by the Book: Food in Literature and Culture*, ed. by Mary Anne C. Schofield (Bowling Green, OH: Bowling Green State University Press, 1989); Carolyn Daniel, *Voracious Children: Who Eats Whom in Children's Literature* (New York: Routledge, 2004); Robert Appelbaum, *Aguecheek's Beef, Belch's Hiccup, and Other Gastronomical Interjections: Literature, Culture, and Food Among the Early Moderns* (Chicago, IL: University of Chicago Press, 2006); Geert Jan Van Gelder, *God's Banquet: Food in Classical Arabic Literature* (New York: Columbia University Press, 2000); Tomoko Aoyama, *Reading Food in Modern Japanese Literature* (Honolulu: University of Hawaii Press, 2008); Susan Skubal, *Word of Mouth: Food and Fiction after Freud* (New York: Routledge, 2002); Paul Vlitos, *Eating and Identity in Postcolonial Fiction: Consuming Passions, Unpalatable Truths* (Cham: Springer International Publishing, 2018); and Nicola Humble, *The Literature of Food: An Introduction from 1830 to Present* (Oxford: Berg Publishers, 2020).

26 Edited volumes of literary approaches to Food Studies often are separate from interdisciplinary collections: see *The Routledge Companion to Food and Literature*, ed. by Lorna Piatti-Farnell and Donna Lee Brian (New York: Routledge, 2018); *The Cambridge Companion to Literature and Food*, ed. by J. Michelle Coghlan (Cambridge: Cambridge University Press, 2020); *Food and Literature*, ed. by Gitanjali G. Shahani

(Cambridge: Cambridge University Press, 2018); and Amy L. Tigner and Allison Carruth, *Literature and Food Studies* (Abingdon: Routledge, 2018).
27 See *Hunger-Proof Cities: Sustainable Urban Food Systems*, ed. by Mustafa Koc and others (Ottawa: International Development Research Centre, 1999); Michael R. Wilson, *Hunger: Food Insecurity in America* (New York: Rosen Publishing Group, 2012); *World Hunger Series: Hunger and Markets* (UK and USA: Earthscan, 2009); John R. Butterly and Jack Shepherd, *Hunger: The Biology and Politics of Starvation* (Lebanon, NH: University Press of New England, 2010); and Alex de Waal, *Famine Crimes: Politics and the Disaster Relief Industry in Africa* (Bloomington: Indiana University Press, 2002).
28 Isabelle Mauret, *Writing Size Zero: Figuring Anorexia in Contemporary World Literatures* (Brussels: P.I.E. Peter Lang, 2007), p. 15.
29 Donnalee Frega, *Speaking in Hunger: Gender, Discourse and Consumption in Clarissa* (Columbia: University of South Carolina Press, 1998), p. 1.
30 Claude Lévi-Strauss, *The Origins of Table Manners: Introduction to a Science of Mythology*, trans. by D. Weightman and J. Weightman (New York: Harper & Row), p. 495.
31 Mary Douglas, 'Deciphering a Meal', *Myth, Symbol and Culture*, 101.1 (1972), 61–81 (p. 64).
32 Roland Barthes, 'Toward a Psychosociology of Contemporary Food Consumption', in *Food and Culture*, ed. by Counihan and Estrik, 23–30 (p. 29).
33 Elspeth Probyn, *Carnal Appetites: FoodSexIdentites* (London: Routledge, 2000), p. 1.
34 Pierre Bourdieu, *Distinction: A Social Critique of the Judgement of Taste*, trans. by Richard Nice (Oxon: Routledge, 2010), p. 191.
35 Deane W. Curtin, 'Food/Body/Person', in *Cooking, Eating, Thinking: Transformative Philosophies of Food*, ed. by Deane Curtin and Lisa M. Heldke (Bloomington: Indiana University Press, 1992), 3–22 (p. 4.).
36 Judith Butler, *Gender Trouble: Feminism and the Subversion of Identity* (London: Routledge, 1990); Michel Foucault, *Discipline and Punishment: The Birth of the Prison* (New York: Pantheon Books, 1977).
37 Sarah Salih, 'On Judith Butler and Performativity', in *Sexualities and Communication in Everyday Life: A Reader* (Thousand Oaks, CA: SAGE Publications, 2007), 55–68 (p. 55); Butler, *Gender Trouble* (1990).
38 Loïc Wacquant, 'Habitus', in *International Encyclopedia of Economic Sociology*, ed. by Jens Beckert and Milan Zafirovski (Abingdon: Routledge, 2006), p. 316.
39 Bourdieu, p. 56.
40 Ibid., p. 6.
41 Maud Ellmann considers the process through a psychoanalytic framework, describing the act of eating as the primal drive that infants initially experience as the method through which the ego comes into being, outranking even the mechanisms of sexuality as the means through which a self-aware subject comes into existence. 'Since sexuality originates in eating, it is always haunted by the imagery of ingestion, having neither an object nor territory proper to itself. Yet eating, in its turn, exceeds the biological demand for nourishment, for it expresses the desire to possess the object unconditionally [...] Most important, it is by eating that the infant establishes his body as his own, distinguishing its outside from its inside.' Maud Ellmann, *Hunger Artists* (London: Virago Press, 1993), p. 39.
42 Ibid., p. 30.
43 Bourdieu, p. 4.
44 Richard Wilk, '"Real Belizean Food": Building Local Identity in the Transnational Caribbean', in *Food and Culture*, ed. by Counihan and Van Estrik, p. 324.
45 Steven Shapin, '"You Are What You Eat": Historical Changes in Ideas about Food and Identity', *Historical Research*, 87.237 (2014), 377–393 (p. 391).
46 Bourdieu, p. 188.
47 Probyn, p. 29.

48. Michel Foucault, *Michel Foucault: The Essential Works, Power, Volume 3*, ed. by Colin Gordon (London: Penguin Press, 2000), p. 98.
49. Hastorf, p. 278.
50. Probyn, pp. 31–32.
51. This lecture was first published in *Foucault Studies* 9 (2010), 71–88.
52. Chloë Taylor, 'Foucault and the Ethics of Eating', in *Foucault and Animals*, ed. by Matthew Chrulew and Dinesh Joseph Wadiwel (Boston, MA: Brill, 2016), 317–338 (p. 318).
53. Hastorf, p. 278.
54. Norbert Elias, *The Civilising Process* (New York: Urizen Books, 1978).
55. Susan Bordo, *The Flight to Objectivity*.
56. Sidney Mintz, *Tasting Food, Tasting Freedom* (Boston, MA: Beacon Press, 1996).
57. Hastorf, p. 281.
58. Pasi Falk, *The Consuming Body* (Thousand Oaks, CA: SAGE Publications, 1994).
59. Gregory J. Seigworth and Melissa Gregg, 'An Inventory of Shimmers', in *The Affect Studies Reader*, ed. by Gregory J. Seigworth and Melissa Gregg (Durham, NC: Duke University Press, 2010), p. 3.
60. Stuart Hall, 'Cultural Identity and Cinematic Representation', *Framework: The Journal of Cinema and Media*, 36 (1989), 86–81 (p. 80).
61. Ben Highmore, 'Bitter After Taste: Affect, Food, and Social Aesthetics', in *The Affect Studies Reader*, ed. by Seigworth and Gregg (2010), p. 120.
62. Ibid., pp. 119–120.
63. Hastorf, p. 274.
64. Neel Ahuja, 'Colonialism', in *Gender: Matter*, ed. by Stacey Alamio, Macmillan Interdisciplinary Handbooks (Farmington Hills, MI: Gale, 2017), 237–251 (p. 245).
65. Roxanne Wheeler, *The Complexion of Race: Categories of Difference in Eighteenth-Century British Culture* (Philadelphia: University of Pennsylvania Press, 2000), p. 21.
66. Juliet Shields, *Nation and Migration: The Making of British Atlantic Literature, 1765–1835* (New York: Oxford University Press, 2016), p. 68.
67. Wheeler, p. 6.
68. Catherine Hall, 'Men and Their Histories: Civilizing Subjects', *History Workshop Journal*, 52 (2001), 49–66 (p. 52).
69. Stuart Hall, 'Conclusion: The Multi-cultural Question', in *Un/settled Multi-culturalisms: Diasporas, Entanglements, 'Transruptions'*, ed. by Barnor Hesse (London: Zed Books, 2000), p. 223.
70. Hall, 'Men and Their Histories', p. 52.
71. Shapin, p. 384.
72. Parama Roy, *Alimentary Tracts: Appetites, Aversions, and the Postcolonial* (Durham, NC: Duke University Press: 2010), p. 6.
73. Isabella Beeton, *How to Dine, Dinners, and Dining, with Bills of Fare for All the Year to Please Everybody* (London: Ward, Lock, and Tyler, 1866), p. 1.
74. Lucy Dow, 'Food, Race and the Nation in Eighteenth and Nineteenth Century British Printed Cookery Books', at Biannual Northern Postcolonial Network Symposium (York St John University, York, UK, 2017), p. 1.
75. Ulrike Thomas, 'Body and Soul: From Tension to Bifurcation', in *A Cultural History of Food in the Age of Empire*, ed. by Martin Bruegel (London: Bloomsbury, 2012), 165–180 (p. 167).
76. Jean Anthelme Brillat-Savarin, *The Physiology of Taste: Or, Transcendental Gastronomy* (Boston, MA: Lindsay & Blakiston, 1854), p. 25.
77. Thomas, p. 169.
78. Edward Said, *Orientalism* (London: Penguin, 2003), pp. 119–120.
79. Sidney W. Mintz, *Sweetness and Power: The Place of Sugar in Modern History* (New York: Penguin Books, 1985), p. 211.
80. Roy, *Alimentary Tracts*, p. 7.

81 Hall, *Men and Their Histories*, p. 52.
82 Ahuja, p. 242.
83 Franz Fanon, *Black Skin, White Masks*, trans. by Charles Lam Markmann (London: Pluto Press, 2008), p. 9.
84 Michelle Pfeifer, '*Becoming Flesh:* Refugee Hunger and Embodiments of Refusal in German Necropolitcal Spaces', *Citizenship Studies*, 22.5 (2018), 459–474 (p. 462).
85 Achille Mbembe, 'Necropoltics', *Public Culture*, 15.1 (2003), 11–40 (p. 40).
86 Vernon, p. 70.

Bibliography

Ahuja, Neel, 'Colonialism', in *Gender. Matter*, ed. by Stacey Alaimo, Macmillan Interdisciplinary Handbooks (Farmington Hills, MI: Gale, 2017), 237–251

Albala, Ken, ed., *Routledge International Handbook of Food Studies* (London: Routledge, 1996)

Allhoff, Fritz and Monroe, David, eds., *Food and Philosophy: Eat, Think and Be Merry* (Oxford and Victoria: Blackwell, 2007)

Aoyama, Tomoko, *Reading Food in Modern Japanese Literature* (Honolulu: University of Hawaii Press, 2008)

Appelbaum, Robert, *Aguecheek's Beef, Belch's Hiccup, and Other Gastronomical Interjections: Literature, Culture, and Food Among the Early Moderns* (Chicago, IL: University of Chicago Press, 2006)

Ashcroft, Bill, *Post-Colonial Transformation* (London: Routledge, 2001)

Ashcroft, Bill and others, eds., *The Post-Colonial Studies Reader: Second Edition* (London: Routledge, 2006)

Ashley, Bob, and others, eds., *Food and Cultural Studies* (London: Routledge, 2004)

Bahri, Deepika, 'Postcolonial Hungers', in *Food and Literature*, ed. by Gitanjali G. Shahani, Cambridge Critical Concepts (Cambridge: Cambridge University Press, 2018), 335–352

Barthes, Roland, 'Toward a Psychosociology of Contemporary Food Consumption', in *Food and Culture: A Reader*, 3rd Edition, ed. by Carole Counihan and Penny Van Estrik (New York: Routledge, 2012), 23–23

Beeton, Isabella, *How to Dine, Dinners, and Dining, with Bills of Fare for All the Year to Please Everybody* (London: Ward, Lock, and Tyler, 1866)

Bordo, Susan, *The Flight to Objectivity: Essays on Cartesianism and Culture* (Albany, NY: SUNY Press, 1987)

Bordo, Susan, *Unbearable Weight: Feminism, Western Culture and the Body* (Berkeley: University of California Press, 1993)

Bourdieu, Pierre, trans. by Richard Nice, *Distinction: A Social Critique of the Judgement of Taste* (Oxon: Routledge, 2010)

Bower, Anne L., ed., *Reel Food: Essays on Food and Film* (London and New York: Routledge, 2004)

Boyce, Charlotte and Fitzpatrick, Joan, *A History of Food and Literature* (London: Routledge, 2017)

Brillat-Savarin, Jean Anthelme, *The Physiology of Taste: Or, Transcendental Gastronomy* (Boston, MA: Lindsay & Blakiston, 1854)

Butler, Judith, *Gender Trouble: Feminism and the Subversion of Identity* (London: Routledge, 1990)

Butterly, John R. and Shepherd, Jack, *Hunger: The Biology and Politics of Starvation* (Lebanon, NH: University Press of New England, 2010)

Caplan, Pat, ed., *Food, Health and Identity* (Oxon and New York: Routledge, 1997)
Carney, Judith A., *Black Rice: The African Origins of Rice Cultivation in the Americas* (Cambridge, MA: Harvard University Press, 2001)
Carruth, Allison and Tigner, Amy L., *Literature and Food Studies* (Abingdon: Routledge, 2018)
Counihan, Carole and Van Estrik, Penny, eds., *Food and Culture: A Reader*, 3rd Edition (New York: Routledge, 2012)
Counihan, Carole and Van Estrik, Penny, eds., *Food and Culture: A Reader*, 4th Edition (New York: Routledge, 2018)
Counihan, Carole and Van Estrik, Penny, eds., 'Why Food? Why Culture? Why Now? Introduction to the Third Edition', in *Food and Culture: A Reader*, 3rd Edition, ed. by Carole Counihan and Penny Van Estrik (New York: Routledge, 2012), 1–15
Curtin, Deane W.,, 'Food/Body/Person' in *Cooking, Eating, Thinking: Transformative Philosophies of Food*, ed. by Deane W. Curtin and Lisa M. Heldke (Bloomington: Indiana University Press, 1992), 3–22
Curtin, Deane W. and Heldke, Lisa, eds., *Cooking, Eating, Thinking: Transformative Philosophies of Food* (Bloomington: Indiana University Press, 1992)
Daniel, Carolyn, *Voracious Children: Who Eats Whom in Children's Literature* (New York: Routledge, 2004)
de Waal, Alex, *Famine Crimes: Politics and the Disaster Relief Industry in Africa* (Bloomington: Indiana University Press, 2002)
Döring, Tobias and others, eds., *Eating Culture: The Poetics and Politics of Food* (Heidelberg: C. Winter, 2003)
Douglas, Mary, 'Deciphering a Meal', *Myth, Symbol and Culture*, 101. 1 (1972), 61–81
Dow, Lucy, 'Food, Race and the Nation in Eighteenth and Nineteenth Century British Printed Cookery Books', at Biannual Northern Postcolonial Network Symposium (York St John University, York, UK, 2017)
Duffet, Rachel, *The Stomach for Fighting: Food and the Soldiers of the Great War* (Manchester: Manchester University Press, 2012)
Earthscan, *World Hunger Series: Hunger and Markets* (UK and USA: Earthscan, 2009)
Elam, J. Daniel, 'Anticolonialism', *Global South Studies*, 2017, https://globalsouthstudies.as.virginia.edu/key-concepts/anticolonialism [accessed 6 September 2020]
Elias, Norbert, *The Civilising Process* (New York: Urizen Books, 1978)
Ellmann, Maud, *The Hunger Artists: Starving, Writing and Imprisonment* (London: Virago Press, 1993)
Falk, Pasi, *The Consuming Body* (Thousand Oaks, CA: SAGE Publications, 1994)
Fanon, Franz, *Black Skin, White Masks*, trans. by Charles Lam Markmann (London: Pluto Press, 2008)
Foucault, Michel, *Discipline and Punishment: The Birth of the Prison* (New York: Pantheon Books, 1977)
Foucault, Michel, *Foucault Studies*, 9 (2010), 71–88
Foucault, Michel, *Michel Foucault: The Essential Works, Power*, Volume 3, ed. by Colin Gordon (London: Penguin Press, 2000)
Franck, Karen A., ed., *Food and Architecture* (Chichester: John Wiley & Sons, 2003)
Freeman, Paul, ed., *Food: The History of Taste* (Berkeley: University of California Press, 2007)
Frega, Donnalee, *Speaking in Hunger: Gender, Discourse and Consumption in Clarissa* (Columbia: University of South Carolina Press, 1998)

Gandhi, Leela, *Postcolonial Theory: A Critical Introduction* (Edinburgh: Edinburgh University Press, 1998)

Hall, Catherine, 'Men and Their Histories: Civilizing Subjects', *History Workshop Journal*, 52 (2001), 49–66

Hall, Stuart, 'Conclusion: The Multi-cultural Question', in *Un/settled Multiculturalisms: Diasporas, Entanglements, 'Transruptions'*, ed. by Barnor Hesse (London: Zed Books, 2000)

Hall, Stuart, 'Cultural Identity and Cinematic Representation', *Framework: The Journal of Cinema and Media*, 36 (1989), 68–81

Harrison, Olivia C., 'For a Transcolonial Reading of the Contemporary Algerian Novel', *Contemporary French and Francophone Studies*, 20. 1 (2016), 102–110

Hastorf, Christine A., *The Social Archaeology of Food: Thinking about Eating from Prehistory to the Present* (New York:Cambridge University Press, 2017)

Highmore, Ben, 'Bitter After Taste: Affect, Food, and Social Aesthetics', in *The Affect Studies Reader*, ed. by Gregory J. Seigworth and Melissa Gregg (Durham, NC: Duke University Press, 2010), 119–120

Humble, Nicola. *The Literature of Food: An Introduction from 1830 to Present* (Oxford: Berg Publishers, 2020)

Inness, Sherrie A., ed., *Kitchen Culture in America: Popular Representations of Food, Gender, and Race* (Philadelphia: University of Pennsylvania Press, 2001)

Ji-Song Ku, Robert and others, eds., *Eating Asian America: A Food Studies Reader* (New York: New York University Press, 2013)

Keller, James R., *Food, Film and Culture: A Genre Study* (California: McFarland and Company, 2006)

Koc, Mustafa and others, eds., *Hunger-Proof Cities: Sustainable Urban Food Systems* (Ottawa: International Development Research Centre, 1999)

Korsmeyer, Carolyn, *Making Sense of Taste: Food and Philosophy* (New York: Cornell University Press, 1999)

Lazarus, Neil, *The Postcolonial Unconscious* (Cambridge: Cambridge University Press, 2001)

Lévi-Strauss, Claude, *The Origins of Table Manners: Introduction to a Science of Mythology*, trans. by D. Weightman and J. Weightman (New York: Harper & Row)

Lionnet, Françoise and Shumei, Shi. *Minor Transnationalism* (Durham, NC: Duke University Press, 2005)

Loomba, Ania, *Colonialism/Postcolonialism: Second Edition* (London: Routledge, 2005)

Mauret, Isabelle, *Writing Size Zero: Figuring Anorexia in Contemporary World Literatures* (Brussels: P.I.E. Peter Lang, 2007)

Mbembe, Achille, 'Necropoltics', *Public Culture*, 15. 1 (2003), 11–40

Mintz, Sidney W., *Sweetness and Power: The Place of Sugar in Modern History* (New York: Penguin Books, 1985)

Mintz, Sidney, *Tasting Food, Tasting Freedom* (Boston, MA: Beacon Press, 1996)

Montanari, Massimo, *Food Is Culture* (New York: Columbia University Press, 2005)

Moore-Gilbert, B.J., *Postcolonial Theory: Contexts, Practices, Politics* (London: Verso, 1997)

Pfeifer, Michelle, 'Becoming Flesh: Refugee Hunger and Embodiments of Refusal in German Necropolitcal Spaces', *Citizenship Studies*, 22. 5 (2018), 459–474

Piatti-Farnell, Lorna and Brian, Donna L., eds., *The Routledge Companion to Food and Literature* (New York: Routledge, 2018)

Probyn, Elspeth, *Carnal Appetites: FoodSexIdentities* (London: Routledge, 2000)

Roy, Parama, *Alimentary Tracts: Appetites, Aversions, and the Postcolonial* (Durham, NC: Duke University Press, 2010)

Roy, Parama, 'Postcolonial Tastes', in *The Cambridge Companion to Literature and Food*, ed. by Michelle J. Coghlan, Cambridge Companions to Literature (Cambridge: Cambridge University Press, 2020), 161–181

Said, Edward, *Orientalism* (London: Penguin, 2003)

Salih, Sarah, 'On Judith Butler and Performativity', in *Sexualities and Communication in Everyday Life: A Reader* (Thousand Oaks, CA: SAGE Publications, 2007), 55–68

Schofield, Mary Anne C., ed., *Cooking by the Book: Food in Literature and Culture* (Bowling Green, OH: Bowling Green State University Press, 1989)

Seigworth, Gregory J. and Gregg, Melissa, 'An Inventory of Shimmers', in *The Affect Studies Reader*, ed. by Gregory J. Seigworth and Melissa Gregg (Durham, NC: Duke University Press, 2010), p. 3

Shahani, Gitanjali G., ed., *Food and Literature*. Cambridge Critical Concepts (Cambridge: Cambridge University Press, 2018)

Shapin, Steven, '"You Are What You Eat": Historical Changes in Ideas about Food and Identity', *Historical Research*, 87. 237 (2014), 377–393

Shields, Juliet, *Nation and Migration: The Making of British Atlantic Literature, 1765–1835* (New York: Oxford University Press, 2016)

Simmons, Marlon and Sefa Dei, George J., 'Reframing Anti-Colonial Theory for the Diasporic Context', *Postcolonial Directions in Education*, 1. 1 (2012), 67–99

Skubal, Susan, *Word of Mouth: Food and Fiction after Freud* (New York: Routledge, 2002)

Taylor, Chloë, 'Foucault and the Ethics of Eating', in *Foucault and Animals*, ed. by Matthew Chrulew and Dinesh Joseph Wadiwel (Boston, MA: Brill, 2016), 317–338

Telfer, Elizabeth, *Food for Thought: Philosophy and Food* (London: Routledge, 1996)

Thomas, Ulrike, 'Body and Soul: From Tension to Bifurcation', in *A Cultural History of Food in the Age of Empire*, ed. by Martin Bruegel (London: Bloomsbury, 2012), 165–180

Toussaint-Samat, Maguelonne, *The History of Food* (West Sussex: Blackwell, 2009)

Van Gelder, Geert Jan, *God's Banquet: Food in Classical Arabic Literature* (New York: Columbia University Press, 2000)

Vernon, James, *Hunger: A Modern History* (Cambridge, MA: Harvard University Press, 2007)

Vlitos, Paul, *Eating and Identity in Postcolonial Fiction: Consuming Passions, Unpalatable Truths* (Cham: Springer International Publishing, 2018)

Wacquant, Loïc, 'Habitus', in *International Encyclopedia of Economic Sociology*, ed. by Jens Beckert and Milan Zafirovski (Abingdon:Routledge, 2006)

Watson, James L. and Caldwell, Melissa L., eds., *The Cultural Politics of Food and Eating: A Reader* (Malden: Blackwell, 2000)

Wheeler, Roxanne, *The Complexion of Race: Categories of Difference in Eighteenth-Century British Culture* (Philadelphia: University of Pennsylvania Press, 2000), pp. 6–21.

Wilk, Richard, ' "Real Belizean Food": Building Local Identity in the Transnational Caribbean', in *Food and Culture: A Reader*, 3rd Edition, ed. by Carole Counihan and Penny Van Estrik (New York: Routledge, 2012)

Wilson, Michael R., *Hunger: Food Insecurity in America* (New York: Rosen Publishing Group, 2012)

Wisker, Gina, *Key Concepts in Postcolonial Literature* (Basingstoke: Palgrave Macmillan, 2007)

Yasmeen, Giséle, *Bangkok's Foodscape* (Bangkok: White Lotus Press, 2006)

2 (Post)colonial Foodways and Transhistorical Hungers in Kiran Desai's *The Inheritance of Loss*

Kiran Desai's 2006 novel *The Inheritance of Loss* is a sprawling, transcontinental narrative that reaches back into an imperial past in order to reflect upon and critique contemporary global power relations, the politics of economic and cultural globalization, and the genealogy of contemporary race relations. The novel's plot takes place in America, England, and India. Food plays a major symbolic role in the book and moves along several different significatory vectors. Considerations of history, the body, gender, and economics are explored and focalized through the alimentary. My aim in this chapter is to lay the basic foundations for understanding many of the themes, forms, and contexts of food and hunger that interweave throughout the body of this book. Although eating is a ubiquitous and mundane process, consumption demarcates and symbolizes complex and specific cultural processes and modalities. Eating is a system of signification that informs and is informed by social relationships within a community. This chapter establishes how eating is a dynamic process that is akin to complex forms of communication. Food and eating, like language, operate dialogically between subjects and circulate within culture and society. Eating, refusing to eat, eating preferences, eating practices, and rituals of eating all help to define the shape, meaning, and boundaries of bodies and identities.

Eating is a continuous cultural process that contributes to the formation of subjects within a socio-cultural framework. The forms of eating and consumption are technologies of the body; that is, they are the means through which the body is affected, affects itself, and acquires meaning. These processes are imbued with signification through their insertion into a vast culturally encoded matrix of power. Reading the body as the location of subjectivity, eating becomes a constitutive process that intervenes and affects at the level of the material and psychological self. It distinguishes somatic identity from others, or in some instances it may produce affiliations. In the western somatic ontology of Cartesianism, the body acquires meaning by incorporating desired food, and consequently desired signification, into the body and defining the desirability of this inclusion against that which is rejected and ejected from the site of the body. Consequently, the process of eating – although seemingly a process of separation and designation – binds subjects intimately with one another in a vast system of identity creation that relies on a notion of difference. Eating is a dynamic process

DOI: -2

of creating the self, as well as a rejection of unwanted traits, associations, and meanings which are then siphoned off on to the bodies and identities of others; this process of 'Othering' is how the subject comes into being within a modern alimentary context.

This chapter defines and explores these alimentary relational dynamics through an analysis of Kiran Desai's Booker Prize–winning novel *The Inheritance of Loss*. This novel is aptly positioned to explore the forms of consumption examined within this study, while also considering the historical contexts that give meaning and context to these forms. I posit that the novel's historical backdrop of food deprivation is tied up with its imperial past and explore how this history repeats in the novel's present through various alimentary relationships. Ideological patterns of eating practices and food values in the novel are responsive to colonial history, and at the same time they are inevitably reconstructed according to present social, cultural, and economic forces. I also consider the first hunger striker examined in this study and explore how this body attempts to reconfigure established meanings of identity, power, and history as they are presented in Desai's novel through its alimentary intervention. *The Inheritance of Loss* stages its narrative in several different historical and geopolitical locations. Modernity and globalization are taken to task in this novel, challenging discourses of development that present a teleology of the world and its history as moving towards a narrowly defined notion of progress. Desai's novel considers 'the ways in which modernity and globalization both rely upon and, in many ways, replicate the same imperial and colonial processes that so many "positive-minded" modern Western thinkers would like to consider world systems of the past.'[1]

The First Hunger Striker: Nimi

The Inheritance of Loss contains an ensemble cast: no single character takes centre stage. Its main setting is the Indian hill-track town of Kalimpong, characterized as a liminal space 'where India blurred into Bhutan and Sikkim.'[2] This community is constituted of several different ethnic groups – Nepalese, Gurkha, Bengali, and Indian – all of whom have distinct national allegiances. The main protagonists of the novel embody and reflect a sense of displacement in that they feel out of place in the melting pot that is Kalimpong. They identify with another place entirely, a place removed from the ethnic groups that comprise the town. The main characters are Sai, educated in an anglicized Catholic school in India, and the judge (formerly known as Jemu), her imposing grandfather and a product of an English education. They are diametrically positioned against their domestic help, who are local working-class Indians. The cook is a hapless but good-natured man who obsequiously caters to the judge's every whim but is treated with loathing in return. The cook's son, Biju, departs for America to find a better life, but only finds the seedy underbelly of alien-immigrant life. Subsidiary characters include sisters Lola and Noni and best friends Uncle Potty and Father Booty, neighbours of Sai and the judge. The novel follows the various characters' exploits in an episodic manner; the narrative is non-linear and includes a number of flashbacks, combined with a main plot that moves forward in a more direct manner. This main story culminates in a brief insurrection in Kalimpong by the Gurkha National Liberation Front (GNLF), a group of young revolutionaries fighting for Indian-Nepali rights.

Nimi, the judge's wife, is a relatively minor character in Desai's novel and does not utter a single line of dialogue throughout the entirety of the narrative – but in this chapter she is examined as a hunger protester. She only features in a few scenes, but her actions are vital towards establishing the focus of this project, and the alimentary logic of the hunger strike itself. Although Nimi is a figure of silence, I contend that the self-imposed hunger she chooses in the novel is an act of protest. Reading Nimi through the dietary technique of self-abnegation, it is possible to establish a theoretical frame for the structure of the hunger strike. Her hunger strike makes sense when situated within a context of food and power as it is contextualized through the complex politics of colonial and postcolonial identity. It is a response to the colonial ideologies as symbolized by food objects and is an effort to shore up her vulnerable status as a female subaltern figure within Desai's narrative. I will explore the connection made between food and language in her hunger strike, a vital symbolic relationship that has repercussions for her strike, and for the other hunger strikers in this project. Nimi is a useful point of departure, as her protest is brief but contains all the theoretical components of the hunger strike as I imagine it in this study. The sprawling nature of Desai's text provides a productive and detailed matrix within which to situate the hunger strike and contains useful readings of racial and alimentary politics that can effectively contextualize the protest.

There are two theories of the body that are relevant for understanding the hunger-striking body in this project. Although theorized separately, these models occur within cultures concurrently, flow from one into another, and can operate on and in the body simultaneously. Both are important for understanding the hunger strike. They are the 'open' body and the 'closed' body. The 'permeable' body is a subset of the open body model. The open body is commonly associated with non-western, primitive societies.[3] In this formulation, the body is open and flexible. There is little boundary between self and the world. The self is eaten into the world (made up of other objects and subjects) and the world is taken into the self when it consumes. This understanding is relevant for the hunger striker as their hunger is conceived as symbolizing a wider social protest within the body of the individual. Individual stands for whole, while the whole finds collective expression in the individual. The second model – the closed body – implies a modern Cartesian body: one that is bounded and non-permeable: 'Controlling the boundaries at an individual level implies a strengthened control over the flows in and out of the body. In other words, the body becomes more "closed" in its relationship to the objects and subjects of the outside world.'[4] The closed body model assumes that what is ingested is integrated into the self; thus, policing the boundaries of the body is important for self-control and self-actualization. This model is vital in understanding why the hunger striker chooses to close off the boundaries of the body to food – as food acts as the symbol for wider discourses of power and colonial ideology, as we will see with Nimi's hunger. The assimilative closed body contrasts with the more 'open,' communicative body. In this model, the act of eating is a complex system of communication whereby substantial meanings are taken into the site of the self – the body – in a dynamic process of becoming. This body is 'created through interpersonal relations and engagements with things in the environment.'[5] This 'permeable' body can be understood as an open body model, but unlike Bourdieu's imagining of the constitutive body being marked by a system of difference in tastes, the open body can be created through exchange and sharing. There is no easily traceable genealogy between these models of the body. Although the closed body is more commonly associated with modern western societies, and the open with more communalistic/pre-modern non-western communities, neither model has vanished completely in the contemporary moment.[6] The postcolonial hunger strike is a nexus for these competing understandings and is so because of the competing discourses that operate on and through the colonial/postcolonial body, situated within the hybridized cultural economy of colonialism.

Nimi's hunger strike is a response to colonial ideology as it is forced upon her by her husband the judge – a main character in Desai's novel. In the novel, colonial ideology is transmitted through the object of food and uses the body as the site to express its sovereignty. A self-imposed hunger strike is the logical strategy through which to stage a revolt. Contained within Nimi's food abnegation is an attempt to police and defend the boundaries of her body from the judge's attempt to civilize her, particularly through processes of westernization. The judge enacts a regime of biopolitical control over Nimi's body as a means of civilizing his own perceived colonial inferiority. Nimi becomes the judge's double – or at least the locus of his projections – and his policing of her body imitates and reaffirms his own self-discipline. The gender imbalance that is embedded in their marriage from its start, as well as the deepening cultural differences engendered by the judge's migration to the west, facilitates the judge's transference of his own symbolic hungers – for acceptance in the metropole, for racial self-negation – on to Nimi, who then manifests these hungers in a more literal fashion. As this particular hunger strike has as much to do with the judge himself as it does with his wife, the first step of analysis should be a detailed interrogation of the judge, particularly with regard to his identity as it relates to the alimentary, his national identity and sense of belonging.

The judge's fraught relationship with himself is chronicled over the course of the novel and from the very first instance is articulated through the medium of food and consumption. It begins with his departure from India to England to join the Indian Civil Service to become a member of the Indian sector of the colonial administration. In the scene of his departure from his village of Piphit to Cambridge, he is named in the narration simply as 'Jemu.' Upon his return from England, Jemu is transformed into 'the judge' – an altered version of his former self. Desai presents this change in moniker to demarcate his transformation from an uncomplicated subject to a complex colonial/postcolonial subject.[7] Before boarding the vessel that will take him to England, his father gives him a ceremonial coconut to throw into the waves as a blessing for his new life. It is the first instance in which Jemu feels the sense of shame that comes to dominate his personality throughout the course of the novel and, subsequently, his life:

> Jemubhai looked at his father, a barely educated man venturing where he should not be, and the love in Jemubhai's heart mingled with pity, the pity with shame [...] He didn't throw the coconut and he didn't cry. Never again would he know love for a human being that wasn't adulterated by another, contradictory emotion.[8]

The rejection of the coconut ceremony signals and symbolizes the beginning of the judge's rejection of his Indianness. The coconut represents a tarnished version of the past for the judge – even in this early state of his border crossing – where once belonging and home were simple and accessible, and communal identity was easily expressed through alimentary rituals.

The situation worsens on the boat, as illustrated by the interaction between the judge and his more middle-class, anglicized cabin mate who 'had grown up in Calcutta composing Latin sonnets in Catullan hendecasyllables.'[9] Jemu's growing awareness of the undesirability of his cultural practices within this new metropolitan context materializes in the visceral disgust towards the judge's native cuisine demonstrated by his cabinmate:

> The cabinmate's nose twitched at Jemu's lump of pickle wrapped in a bundle of puris; onions, green chillies, and salt in a twist of newspaper; a banana that in the course of the journey had been slain by heat [...] Jemu picked up the package, fled to the deck, and threw it overboard. Didn't his mother think of the inappropriateness of her gesture? Undignified love, Indian love, stinking, unaesthetic love.[10]

The Indian meal prepared lovingly by his mother takes on a different meaning within the context of more desirable cultural capital, indicated by Jemu's cabinmate's superior aesthetic signifiers of 'Donegal tweeds'[11] and a fine 'gilded volume'[12] of sonnets. Jemu's cabinmate serves to illustrate the burgeoning illegitimacy of Jemu's identity, engendered by migration, and as this example takes place before his encounter with the metropolitan centre, one can assume that it is also produced by imperial discourse accessible in the colonized space itself. This cultural hierarchy is magnified by encounters with more westernized subjects. As the son of a poor Indian, Jemu has not had the appropriate cultural education, and so his body and mind are not yet disciplined in the 'correct' way. Jemu's cabinmate signals a range of appropriate bodily practices – he consumes the correct texts, he eschews incorrect foods, and his body adheres to the correct sartorial codes.

Throughout his stay in Cambridge, the novel chronicles Jemu's internalization of the inferior status of the colonial Other, as it 'crushed him into a shadow'[13] and his 'mind began to warp; he grew stranger to himself than he was to those around him, found his own skin odd-colored, his own accent peculiar'[14] until eventually 'he felt barely human at all.'[15] His own fractured subjectivity and perceived role in the narrative as illegitimate – as a pretender – is exemplified by his name change. The moniker indicates a severe criticality, rigidity, and a didactic sense of right and wrong. 'He envied the English. He loathed Indians. He worked at being English with the passion of hatred and for what he would become, he would be despised by absolutely everyone, English and Indians, both.'[16] The judge finds himself at the intersection of a double bind. He is educated and encouraged by the metropole to deny and overcome his own inferior racial identity by a system whose underlying logic is that colonial subjects should be civilized. However, this same context apprehends the native body as biologically distinct – more animal than human – and essentially so, defeating any such civilizing endeavours. 'Among colonizing elites – even if they shared a conviction of superiority – tensions often erupted between those who wanted to save souls or civilize natives and those who saw the colonized as objects to be used and discarded at will.'[17] Both theories position the colonized body in an impossible state, one characterized by simultaneous and contradictory self-policing and attempted self-transcendence.

Frantz Fanon provides perhaps the most well-recognized image of this figure of the native in his book *Black Skin, White Masks*. In it, he presents a mimic man. The mimic man is caught between what he aspires towards – a complete transformation into his colonial masters – and what he cannot deny in himself: his cultural/racial identity. Fanon didactically presents the problem of the colonized people. The example given compares the black man with the white, but the theory may be transplanted to other contexts – historical, cultural, and racial. 'All forms of exploitation resemble one another.'[18] Fanon presents his case succinctly: 'There is a fact: White men consider themselves superior to black men. There is another fact: Black men want to prove to white men, at all costs, the richness of their thought, the equal value of their intellect.'[19] This desire to prove themselves to white men stems from an internalization of the material and psychological inequalities that characterize and are perpetuated by imperialism: 'I believe that the fact of the juxtaposition of the white and black races has created a massive psychoexistential complex.'[20] In *The Inheritance of Loss*, this internal battle rages with particular ferocity in the judge. His subjectivity is characterized by a constant state of striving and aspiration but is also understood to be incapacitating, fomenting an imbalance that naturally requires and desires resolution. The desired resolution is, of course, impossible, within the terms outlined by colonial civilizing discourse. The postcolonial subject will never fully achieve the status of the white man, and yet he still cannot fully disavow his own body, its differences, and its histories.

The judge's relationship with his wife, Nimi, begins as innocent attraction, but because of his psychological transformation, it becomes troubled by enmeshment and resentment. Both the former Jemu and the present judge have a relationship with Nimi – one before England, one after – and these two relationships differ in the extreme. Jemu's marriage to the 14-year-old Nimi is arranged shortly before he leaves for Cambridge, to ensure his literal and metaphorical marriage to his culture and roots. Before he leaves, they have a brief but magical encounter as Jemu feels the first stirrings of infatuation for his beautiful young bride. They share an innocent bicycle ride that Jemu remembers fondly when he first leaves India; his sexual attraction toward her is tentative but desirous: 'However in memory of the closeness of female flesh, his penis reached up in the dark and waved about, a simple blind sea creature but refusing to be refused. He found his own organ odd: insistent but cowardly; pleading but pompous.'[21] His body here is described ambiguously – animalistic, forceful but frightened, and somewhat uncontrollable. The phallus metonymically refers to Jemu's body, gesturing towards racial theories that place native bodies nearer to animals than humans. This body and its desires are rendered abject and Other. His indoctrination in England against all things Indian forces him to reassess his initial sexual desire, and where and how to direct it. Upon his return to Piphit, he immediately struggles with his connection to Nimi, as she represents everything he loathes in himself and his countrymen. The revulsion is both a projection of the self onto Nimi and an awareness that what he loathes resides in himself. 'Her presence is "disruptive" because she reminds him of the contradictions that he tries to suppress.'[22] His young wife seems willing to engage with the marriage when the judge returns from England, but his mixed feelings toward her transform it into a union filled with anger, disgust, and violence. His first experience of sexual intimacy becomes a disturbing scene of rape, during which his desire for her is articulated both as ravenous and repulsive. Sexual union becomes an act of destruction, fuelled by the same drive: the hatred of the self, reflected in the Other. Nimi becomes a mirror for the judge's self-loathing, and despite the unsavoury nature of their intercourse,

> he repeated the gutter act again and again. Even in tedium, on and on, a habit he could not stand in himself. This distaste and his persistence made him angrier than ever and any cruelty to her became irresistible. He would teach her the same lessons of loneliness and shame he had learned himself.[23]

Here, desire and disgust become intertwined. Jemu finds himself in the grip of a struggle between the rational restraint he has learned in England and his uncontrollable bodily desires for Nimi. The concurrent hypersexualisation and abjection of the female colonial body here intersects with the colonial savagery associated with the judge's own body. 'The "exotic" woman of color became a sexual object and was no more a proper wife than a prostitute.'[24] The judge's sexual desire is a reminder of the lasciviousness of his own uncontrollable colonial body, while his desire for his wife is structured by imperial fantasies of the Orientalized female native – themselves reinforced by the assumed animalism of the colonial Other.

Over the course of many years of this psychological and physical abuse, Nimi begins to resist. She begins to reject food. This refusal is read as a rejection of her husband's idealized western cultural values. The novel uses food as a means of articulating power, marking cultural boundaries, and eating is read as a means of interpolating subjects – one becomes what one eats. Nimi's food refusal is a rejection of the judge's civilizing efforts. She begins to refuse food, and this deliberate abnegation is accompanied by a simultaneous rejection of words:

> Nimi learned no English, and it was out of stubbornness, the judge thought.
> 'What is this?' he questions her angrily, holding aloft a pear.
> 'What is this?' – pointing at the gravy boat bought in a secondhand shop, sold by a family whose monogram had happily matched, *JPP*, in an extravagance of flourishes. He had bought it secretly and hidden it within another bag, so his painful pretension and his thrift would not be detected. *James Peter Peterson or Jemubhai Popatlal Patel.* IF you please.
> 'What is this?' he asked holding up the bread roll.
> Silence.
> 'If you can't say the word, you can't eat it.'
> More silence.
> He removed it from her plate.[25]

Nimi remains silent and refuses to conform to the judge's standards of dining etiquette and language. '[D]istinctions in food preparation, eating habits, and modes of dining are a crucial axis around which cultures and groups consolidate themselves.'[26] The dinner table is a central site for cultural expression, as well as familial power relations. As Isabella Beeton reminds us: 'Dining is the privilege of civilization.'[27] Deborah Lupton focuses the civilizing principles of the dinner table on to the family community: 'The "family meal" and the "dinner table" are potent symbols, even metonyms, of the family itself.'[28] In the family, power is enshrined at the dinner table, so the judge's oppressive dietary rules are reflective of not only the larger context of colonial ideology but also how this plays out in individual domestic settings.

In this scene, a vital connection is made between food and language.[29] Speaking and learning English become symbolic of Nimi's refusal to succumb to not just her abusive husband but also the colonial discourses that structure his identity and her national context. As Frantz Fanon states in *Black Skin, White Masks*, acceptance of the colonialist's language is the ultimate acceptance of colonial culture: 'Every colonized people – in other words, every people in whose soul an inferiority complex has been created by the death and burial in its local cultural originality – finds itself face to face with the language of the civilizing nation.'[30] Nimi's body is chosen as the site upon which to enact a defence against her husband and all he represents. Maud Ellmann notes in *The Hunger Artists* that 'eating is essential to the "integration" of the self, but this is only if the food is voluntarily ingested. Force-feeding, on the contrary, demolishes the ego.'[31] Nimi's hunger is highly relevant because it associates the ingestion of food with the ingestion of words. The associative relationship between the two is set up as corresponding. '[F]ood is a social and semiotic fact, and language plays a key role in its construction, evaluation, mediation, and definition.'[32] In denying Nimi food at the expense of words, the judge creates a connection between the signifier (the food image) and the signified (the meaning of the food image). In this instance, the signified is markers of civilization, routed through the alimentary discipline of the body via the signifier: food. 'If you can't say the word, you can't eat it.' With this strategy, the judge iterates a belief that language has a defining and creative power over the body as it is symbolized and doubled by ingesting the 'correct' foods: '[for the colonized body] diet was configured from the nineteenth century on as a terrain for encounter, challenge, transformation and consolidation.'[33] Language, and food, contains the capacity to transform the material self – the body. Language/food is exalted as an intellectual and civilizing force that contains the ability to discipline and produce new meanings in and of the body. Nimi's material body is codified by the judge as requiring re-inscription, and with that comes erasure. He imagines the ingestion of civilized, English words – and eating practices – can rewrite Nimi's native body.

In 'Tasty Talk, Expressive Food: An Introduction to The Semiotics of Food-and-Language,' Riley and Cavanaugh offer four ways of distinguishing food and language relationships: (a) language through food: the use of food to communicate emotion, identity, distinction, and social relations; (b) language about food: how to speak about food as a subject matter; (c) language around food: the genres of interaction in the presence of food; and (d) language as food: the understanding of communication as a form of nourishment.[34] For the purposes of examining Nimi's hunger-as-protest, (a) and (d) are the most relevant. Food is used as a means of communicating certain ideologically inflected imperial discourses between subjects, and language is replaced by food items as an alternative form of ingestion – both language and food can communicate meaning and value.

In his article '"Solid Knowledge" and Contradictions in Kiran Desai's *The Inheritance of Loss*', David Spielman states:

> Nimi arouses in the judge ambivalent feelings, in the strictly psychoanalytic sense of the word. He simultaneously desires and rejects his wife, and these feelings are interdependent. His response to this contradiction, as with others, is to suppress it by removing Nimi from his life entirely and forgetting her.[35]

I agree with Spielman's contention that the judge's feelings towards Nimi are ambivalent, and that the desire and revulsion he feels towards her are mutually reinforcing. But I would argue that the judge's method of dealing with this contradiction is far removed from ignoring Nimi, or from deleting her from his life. As is evidenced by his efforts to contingently feed – and then deny – her both words and food, the judge attempts to solve the problem of his own ambivalent identity by resolving his abject feelings as they are projected onto Nimi's body. His effort to get her to conform through this protracted and forced eating practice is an attempt to exorcize that which he finds repulsive within himself. The judge's intended enforcement takes the form of western cultural capital: the English language and western cuisine. Nimi refuses them both, as they represent not only the unwelcome invasion of her body but also the ingestion of the colonial ideology that has twisted her husband into the hateful man he has become. Her hunger strike, and her accompanying silence, demonstrate that food and words share a common logic in that they both involve a speaker and a corresponding interlocutor. Nimi's intent, articulated in her refusal of food, may never be grasped fully by the reader, as her silence precludes any certain interpretation. Her body becomes a text to be read and interpreted by the novel's reader. We cannot be sure if the judge himself is able to decipher the form of Nimi's protest, although we can ascertain the meaning of his alimentary strategies through interpretations of Desai's text.

As a projection of the judge's divided psyche, Nimi attempts to intervene in the parasitic relationship between herself and the judge by erecting a protective field around her body. This is a deployment of the 'closed' body model – the body's boundaries are closed off as a means of articulating individual autonomy and separateness from those around it. However, the permeable body is also relevant here, as the communicative framework of the food-as-language contained in the hunger strike demonstrates that Nimi and the judge's identities are painfully interdependent – not based on a system of difference, but rather one of doubling. When social relations are articulated through food, it is almost impossible to conceive of them as limited to the self. As Spielman suggests above, the judge soon starts to completely ignore Nimi's existence, and even the servants begin to treat her like a ghost. The narrative states, 'She had fallen out of life altogether.'[36] She does almost disappear from the plot from this point onward; she is barely mentioned in the proceeding chapters. However, she remains intact as an idea, a category, a symbol of the disavowed native inferiority that invades the judge's psyche. She is not a person at all, but a discursive boundary within the judge's subjectivity, unbreachable, and immovable to the last. Nimi's eventual exit from the novel is a final bid to control the destiny of her own body: she self-immolates and finally puts a stop to the 'limitless bitterness'[37] she and the judge have for one another. This hunger strike ends in death – as so many do – and so elicits questions about the success of violent somatic protests like these. The body is asserted as a ground for dissent – perhaps the only available tool with which to revolt in especially powerless conditions – but its mortal frailties are ultimately its own undoing.

This hunger strike is an example of how a self-styled biopolitical technique is deployed in response to other forms of biopower – here, specifically, the biopolitics of the colonizer's discourse of power. We as readers have no access to Nimi's interiority, but third-party reading of her body may produce productive politics. As Amanda Machin states in 'Hunger Power: The Embodied Protest of the Political Hunger Strike':

Undertaken by those denied voice it nevertheless can be extremely powerful. It deftly interiorises the violence of the opponent within the body of the protester, affirming and undermining the protest simultaneously. It can be undertaken for highly strategic and rational reasons and yet it is often affective because of the emotional response it provokes [...] In particular, it highlights three political aspects of the hunger strike: 1) the facilitation of non-verbal communication 2) the embodiment of collective identifications 3) the disruption of the dominant order.[38]

Through her suffering, Nimi can experience herself as a political actor with access to some form of agency. Through the active and violent textualization of her body, Nimi can communicate and respond to what is effectively a somatic imprisonment. As mentioned in the quote above, the body can assert itself, thus challenging the dominant norms of legible communication – the privileged forms of discourse available to rational actors – in this case, agents of imperialism. Embodied protest therefore goes some way toward rewriting the archetypes of the sensuous, bodily native. The body is foregrounded as a means of articulation that disrupts the dominant order of rational debate. Of course, contradictions also appear within this form of protest, not least because of the severe physical cost of such an act. The Cartesian body/mind binary upon which the protest is ultimately structured remains intact. In the hunger protest, the will of the mind disciplines the animal desires of the body; therefore, as a technique it replicates the confining racist structures associated with Cartesianism – the 'native' is somatic creature of instinct and uncontrollable desire, and the civilized (in this case always western body of Enlightenment) subject controls somatic urges with the rational control of the mind. But there are ways of reading political value into the hunger protest. A common understanding of hunger strikes is that they are primitive forms of protest (as opposed to more 'civilized' forms such as rational debate), and read through this theory, a hunger-striking body may simply re-entrench the native body further into the Orientalized figure of colonial fantasy – a primitive body protesting in primitive ways. 'Philosophy is masculine and disembodied; food and eating are feminine and always embodied.'[39] Relying on somatic and alimentary protest, Nimi deploys an illegible response, according to western philosophical ideals. However, the hunger strike can be read as a means of asserting not just the body itself, but the *discursive* power of the body:

[T]he body also serves as a political actor; the one who scrawls the text through a self-directed violence. The body is not simply a 'docile object' that is passively conditioned or violently constructed; it also creatively contributes to political protest itself.[40]

The power of speech is not negated; rather, it is transferred in productive ways on to the textual capability of the body. Nimi speaks through her bodily pain and food refusal. This example of self-starvation is a useful starting point in considering the meanings of self-imposed hunger in the colonial/postcolonial context, but the meanings of hunger never exist in a vacuum. The relevance of hunger, particularly in postcolonial writing, must be framed by historical contexts. It is necessary to consider the specific historical events around food, hunger, and food scarcity that contribute to the emergence of food and hunger as loaded symbols in postcolonial literature. The following section considers how the meanings of food – as they engage with and are defined by an economy of imperial power whose effects are felt in both the past and present – are established and then appear and reappear in postcolonial narratives.

Historical Contexts: A History of Hunger

In *Famine: Social Crisis and Historical Change*, a text that brings a historical perspective to famine and starvation, David Arnold outlines some of the drawbacks arising from the lack of interdisciplinary insight into the phenomena of starvation: 'Interestingly enough, it is more often demographers and economists, geographers, anthropologists and political scientists, rather than historians, who have made the running in the recent discussion of famine and who have advanced many of the most challenging theories.'[41] He elucidates in his introduction, saying that historical perspectives on the matter have generally been confined to examining the intricacies of specific famine events rather than attempting to provide an overarching significance of famine. Due to this, he says, it has been 'left, therefore, to the non-historian to speculate upon the more general place of famine in the human experience.'[42] While a number of historical studies about famine have emerged since Arnold's assertion, his claims are useful in a consideration of the disciplinary boundaries that are traversed when considering hunger on a mass scale. Generally, the study of famine has been left to the social scientist, but not merely around the issue of causality. Famine has thus been framed by discipline's discursive trends and aims.[43] The disciplines that fall under the umbrella of social sciences are often encouraged to collect data that might augment policy, and contain an underlying thesis to instigate social change. This contributes to the tone of practical social application. But literature can locate the famine in a more microcosmic register, as well as considering forms of textuality present in the representations of famine. The greater emphasis on different types of epistemologies is perhaps best examined through fiction. As Margaret Kelleher argues, 'perhaps the most significant aspect of a literary famine text is its potential to individuate the crisis.'[44]

Causality has generally been the remit of sociological studies of food insecurity. Arnold's own historical intervention into the discussion of food is largely centred on the notion of power. He states:

> Historically food was one of the principal sinews of power. Its importance was felt at all levels of society, both by those who suffered directly for want of basic sustenance and those whose authority, security and profit

were threatened as the indirect consequence of dearth and mass starvation – from the family to the state, from the peasant households of medieval Europe or Imperial China to the colonial empires of the nineteenth century and the international market economy of the present day. Food was, and continues to be, power in a most basic, tangible and inescapable form [...] Food (and the denial of absence of food that famine entailed) was (and remains even in a relatively secure and secularized society like our own) richly symbolic, a potent and recurrent motif in the semantics of kingship and statecraft, in the language and imagery of religion and culture.[45]

This project is interested in the way 'food was, and continues to be, power in a most basic, tangible and inescapable form,' and explores the function of these powers in Desai's novel. This section interrogates the permutations of power that are constructed through representations of food, hunger, and eating, historically and within the world of the novel itself. To facilitate this analysis of present narrative forms and contexts, I investigate historical narratives of hunger and famine. These historical contexts provide the framework for understanding how power operated and was operationalized with regard to historical hunger and deprivation, with a focus on imperial interventions. In the novel, these histories are articulated through food, eating practices, hungers, and the way characters construct their alimentary identities in relation to each other.

In his book *Late Victorian Holocausts*, Mike Davis puts forward a compelling and convincing causative summation of a series of devastating famines during the late Victorian age. These famines affected massive swathes of Africa and Asia. A total of three devastating global droughts resulted in a total death toll of more than 30 million victims. He calls these subsistence crises 'the secret history of the nineteenth century,'[46] demonstrating how these huge food failures and death tolls have been marginalized by metropolitan famine history writing. The 'secret' he refers to, however, is not merely the elision of historical narratives concerning the famines' existence but also the contextual, causative parameters of the crises. He echoes Arnold's sentiment in asserting that climatic failure fails to provide the whole picture in assessing the mitigating factors behind famine. Although climatic factors were undoubtedly a huge aspect of the food crises,[47] failure in colonial administration exacerbated or even took advantage of them.

Until the nineteenth century, causative theories of hunger and famine had prioritized inevitability. Images of climatic disaster and nations engulfed by overpopulation were the dominant representations associated with instances of both chronic and acute hunger. Meteorological extremes such as floods, tsunamis, droughts, and earthquakes warn of impending shortages of shelter, medicines, resources, aid, and, most importantly, food. The issue of overpopulation provides a somewhat less shocking and more gradual arrival at the same anxieties: the ever-multiplying populations of underdeveloped countries might overtake global resources and the world's capacity to support humankind. Causative narratives of this nature place the onus on a higher power – fate, God – or a collective, and consequently unaccountable, essential 'human nature.' This reasoning deflects attention away from the individual, producing the notion that human beings have little impact on the direction of hunger. This applies to both the social conditions that would allow hunger to exist and the external causes that might trigger a period of food insecurity. This type of thinking is markedly Malthusian in nature, inspired by T. Robert Malthus and his theory on 'preventive' and 'positive' checks in his *Essay on the Principle of Population*. He, along with many before and after him, argued (and have continued to argue) that famines are a product of human populations overutilizing resources and that the natural outcome of such excess is the 'positive' check of famine.[48] He viewed natural disasters similarly – as inevitable and unstoppable events that serve to cap the size of human populations. Despite the fact that ensuing research and efforts to rehistoricize hunger causality have demonstrated that technological advances in food production have kept pace with population growth and so invalidate fears of global overconsumption,[49] populist narratives continue to pin responsibility for food shortage on meteorological factors or other inevitable causes.

Within the academy, contemporary theories on hunger have emerged as an antidote to Malthusian approaches. Popularized by economist Amartya Sen, particularly in his book *Poverty and Famines*, the food-entitlement approach entails an economic assessment of the individual's purchasing power as a means of measuring their ability to secure food – a purchasing power that is subject to supply and demand. As Sen states in the opening line of *Poverty and Famines*: 'Starvation is the characteristic of some people not *having* enough food to eat. It is not the characteristic of there *being* not enough food to eat.'[50] Sen argues that in most food-subsistence emergencies, crises of chronic and acute hunger do not result from the quantity of food but, rather, from its unequal distribution. It is an individual's ability to secure food, or, as Sen termed it, 'food entitlement', that is responsible for his or her lack of food security. And as food entitlements are subject to the fluctuations of a market economy, it is the forces of supply and demand that keep an individual from securing food. Essentially, the method approaches causality through an economic lens, and thus attributes accountability to the man-made market. The consideration of economics within an existing framework of causality that posits the weather or overpopulation as its origin marks a shift in understandings of hunger (on a global or national level) from 'natural' to 'unnatural,' and thus the issue becomes politicized.

This is not to say that climatic factors should be dismissed when theorizing famine. Contemporary models of food insecurity consider climatic, social, and political factors. It may be more accurate to conceptualize famine as having two causative dimensions: famine as event and famine because of pre-existing social structures. It is an event in that it occupies a finite time frame, and it is glaringly obvious when a famine is occurring: as Stephen Devereux states in *Theories of Famine*, 'famine is like insanity, hard to define, but glaring enough when recognized.'[51] At the same time, a famine is also a structure, as they rarely appear out of the blue. Although there may be famines that are caused by specific incidents that are highly localized and unpredictable (such as the blockade of Germany during the Second World War), usually this is not the case. Barring some exceptions, a famine is the shocking aftermath of a 'tipping point' that exacerbates already-existing social weaknesses:

> [A] famine acts as a revealing commentary upon a society's deeper and more enduring difficulties. The proximate cause of a famine might lie in some apparently unpredictable 'natural disaster', like a flood or drought, or in a 'man-made' calamity like a civil war or invasion; but these are often no more than the precipitating factors, intensifying or bringing to the fore a society's inner contradictions and inherent weaknesses, exposing an already extant vulnerability to food shortages and famine.[52]

Climatic parameters, as well as social, political, and economic factors, combine to provide an available set of principles that can be drawn upon to understand famine causation.

While the scholarly debate surrounding food shortage has moved on, public assumptions about food insecurity remain entangled with the notion that food insecurity comes about, quite simply, when there is not enough food to go around, along with other more Orientalist conceptions about the types of nations and races that are more commonly associated with food insecurity. Food security is a marker of Modernity. 'The stereotypical association of hunger and famine with Africa and South Asia (and once with the Soviet Union, China, and Ireland) obscures the extent to which food insecurity is experienced more pervasively.'[53] Daniel Maxwell states:

> For many years experts believed that famines were caused by a shortfall in food availability. Then in 1981 economist/philosopher Amartya Sen published 'Poverty and Famines: An Essay on Entitlement and Deprivation,' which showed that famines actually resulted when food was available but some groups could not access it. Although many people believe today that famines occur mostly in Africa, the deadliest famines of the 20th century were in Europe (Ukraine) and Asia (China). Today we recognize famines happen only with some degree of human complicity.[54]

This illustrates popular beliefs around famine and food shortage, as well as how these ideas intersect with Orientalist beliefs about underdevelopment and where poverty is geopolitically located. It also raises the question as to why these beliefs remain so staunchly held, particularly when the scholarly discussion around food security has moved on:

> Such an understanding of famine as a problem of reconciling this clash of understandings has been and continues to be pervasive among international relief organizations, in spite of the moral force of Sen's argument that famines, at least in the modern period, are caused not by shortfalls in production or the lack of the transportation that would enable the easy

movement of foodstuffs but by powerful social inequalities that he describes as class- and gender-stratified 'entitlements.'[55]

This theoretical gap may be explained by the intersection of Malthusian discourses with neoclassical economic thought, as exemplified in Adam Smith's idea of the 'invisible hand':

> [H]e intends only his own security; and by directing that industry in such a manner as its produce may be of the greatest value, he intends only his own gain, and he is in this, as in many other cases, led by an invisible hand to promote an end which was no part of his intention. Nor is it always the worse for the society that it was no part of it. By pursuing his own interest he frequently promotes that of the society more effectually than when he really intends to promote it. I have never known much good done by those who affected to trade for the public good.[56]

Currently, the term 'invisible hand' is deployed in several different contexts and can mean different things. But it is widely understood as referring to a process by which economic outcomes are produced in a decentralized way, with no explicit, or intentional, agreements between participating actors. The market is 'free' and is controlled by an invisible force that checks itself in order to operate optimally. Adam Smith was a religious man, and the image of an 'invisible hand' evokes the Christian logic of benevolent overseer. Rationality is the key characteristic of this divine force. Free, self-interested subjects operate rationally and receive a commensurate and just reward. The laissez-faire parameters of neoclassical economics determine its adherence to this perceived moral centre. Similarly, 'Malthusian catastrophes' are predicated upon this utopian model of the market because acute events such as famines and meteorological disasters were viewed as necessarily 'checks' on the population. Malthus suggested that a food disasters were natural and appropriate. Both Malthus and Smith shift responsibility from the market to powerful forces that cannot be controlled – and indeed should not, given the moral imperative they signify. As Malthus states:

> It has appeared, that from the inevitable laws of our nature some human beings must suffer from want. These are the unhappy persons who, in the great lottery of life, have drawn a blank [...] no possible sacrifices of the rich, particularly in money, could for any time prevent the distress among the lower members of society, whoever they were.[57]

The suffering of the poor is characterized as necessary and inevitable. This inevitability, combined with a 'rational' market that is guided by moral invisible forces, discourages critical examination into individual actions and human impact in food shortage situations.

The neoliberal discourses of late capitalism resonate with Malthus's theories of food shortage, and thus the idea persists that famines result from food-supply failure. Famine is seen as inevitable, and something for which no individual, collective, or institutional body is accountable. Instead, famines are read in problematic ways. Indeed, famine and food security become something mythical. Parama Roy comments on Malthus's apocalyptic vision on famine: 'What is note-worthy here is a fantasmatic array of images and figures that allegorizes famine as something tremendous, irresistible, and virtually otherworldly, something that outstrips the logic of supply and demand that permeates the analysis of the causes of famine.'[58] This demonstrates a reluctance to see what truly animates food insecurity and instead mythologizes it as an inevitability, worryingly moral in tone, and more often than not racialized. Neoliberal ideas around the competitive yet wholly fair nature of capitalism fuel racist stereotypes around countries and individuals who are commonly associated with food lack. The representation of black and brown bodies starving in Africa and Asia is seen as accurate because their land is less fertile (or their population too fertile, reproducing at alarming rates), their resource management poor, and their citizens neither enterprising nor productive.

Inserted within a system governed by economics, then, food becomes a symbol of power and a sign of global competence. Desai's novel considers the dynamic interplay between popular and received wisdom about hunger as a moral force, and understanding food as a measure of power, accessible or inaccessible by human actors. The novel is overtly critical of globalization and draws distinct historical connections between the present and the imperial world order that precedes it. Food operates in *The Inheritance of Loss* as a symbol that travels from past to present, containing within its meanings histories of imperial domination and racial inequality. In the following section, I trace a genealogy of hunger through a narrative of historical food insecurity in India, and consider how these events contribute to the current world economic order that also determines the value, meaning, and desirability of food – and then in turn the implications this has for postcolonial identities.

In the late eighteenth and nineteenth century, the Indian subcontinent experienced a series of devastating subsistence crises. These were particularly bad in 1876–79 (when an estimated 25 million perished) and 1896–1902 (when an estimated 45 million perished).[59] Although India had faced famines repeatedly in the past, these crises reached a crescendo under British colonial rule, and the continent saw the worst incidences of hunger during this period. In the case of India, the nation's economy had already been shoehorned into a global capitalist economy, whereby a food shortage meant rocketing prices. Thus, traditional defences against crop failure no longer worked. Karl Polyani explains in *The Great Transformation* the impact of capitalism-oriented responses to the late-nineteenth-century droughts:

> In former times small local stores had been held against harvest failure, but these had been now discontinued or swept away into the big market [...] Under the monopolists the situation had been fairly kept in hand with the help of the archaic organization of the countryside, including free

distribution of corn, while under free and equal exchange Indians perished by the millions.[60]

Without these traditional defences, the resulting food shortage could not be tackled in the usual way. This was due to the country's economy being rearranged into a global capitalist framework that had been introduced by the British colonial administration. This capitalist framework was an import and reflected post-industrial European economic paradigms that had accompanied and now defined colonial expansion. Britain's economy enjoyed the benefits of a wide variety of cheap imports, while millions in imperial India died for want of food:

> We are not dealing, in other words, with 'lands of famine' becalmed in stagnant backwaters of world history, but with the fate of tropical humanity at the precise moment (1870–1914) when its labour and products were being dynamically conscripted into a London-centred world economy. Millions died, not outside the 'modern world system,' but in the very process of being forcibly incorporated into its economic and political structures. They died in the golden age of Liberal Capitalism; indeed many were murdered [...] by the theoretical application of the sacred principles of Smith, Bentham and Mill.[61]

This argument claims the British administration wilfully allowed the famine to occur by their uninterrupted export of grain to their own markets. Besides this, the British administration knew well that the food shortages were not wholly climate-dependent, but denied this knowledge in order to grease the wheels of capitalistic expansion:

> From the 1860s, or even earlier, it was generally recognized in India, both by British administrators and Indian nationalists, that the famines were not food shortages per se, but complex economic crises induced by market impacts of drought and crop failure.[62]

These policies created the 'death-worlds' that Achille Mbembe describes in 'Necropolitics.' Building on Foucault's biopower, necropolitics not only refers to how lives of citizens are controlled but accounts also for how deaths of citizens are controlled – or a 'subjugation of life to the power of death.'[63] These Great Victorian Holocausts can be framed as directly or indirectly administered by imperial power, a centralized force whose ideologies determine who has the right to live or die, and in what manner. Under the Viceroy of India – Lord Lytton, 'Victoria's favourite poet'[64] – laissez-faire doctrines were strictly adhered to, and even contemporary British commentators were 'appalled by the speed with which modern markets accelerated rather than relieved the famine.'[65] The images of the Bengal famine that made it to metropole did trigger the beginnings of a more humanitarian approach to famine that, to a certain extent, challenged Malthusian ideals of the self-correcting market and population. However, even these humanitarian efforts were framed by the civilizing discourse of imperialism. As James Vernon writes in *Hunger*: 'It was a narrative that captured the myth of British colonial rule and neatly reconciled the existence of famine with the rhetoric of improvement and the forward march of the civilizing mission in India.'[66]

The historical legacy of this arrangement of colonial economies into dependent relations with more powerful countries is evidenced in the present. The global market economy as it stands today can be traced back to this period of colonial expansion:

> [W]hat we today call the 'third world' (a cold war term) is the outgrowth of income and wealth inequalities – the famous development gap – that were shaped most decisively in the last quarter of the nineteenth century, when the great non-European peasantries were initially integrated into the world economy.[67]

The spread of capitalism via colonialism established a global economy whose historical system of inequality still operates today: socially, economically, and culturally.

> The historically grounded social critique of classical political economy and industrialization that developed in Britain was to some extent foreshadowed by the work of nationalists in Ireland and India. They demonstrated that the human cost also included the millions of lives lost to colonial famines. The economic modernization of Britain, they suggested, depended on the underdevelopment of its colonial economies.[68]

Colonialism fostered global underdevelopment by incorporating other nations into its own economic supply and demand requirements.[69] In *The Political Economy of Underdevelopment*, Amiya Kumar Bagchi traces the development of underdevelopment in India from its colonial roots, starting with the British East India Company, through to India as a state-controlled colony, to post-independence. 'India had in effect made a transition from the demand-constrained stasis of colonial times to the multiply-constrained three-legged race of a neo-colonial, retarded society.'[70] Bagchi draws a direct causative line from imperial India to its present-day problems with poverty and underdevelopment.[71] Theories such as these have expanded to include other ex-colonies. Considering our present globalized economy, critics have expressed a more totalizing theory to explain inequality: 'The colonial era left most countries of the South in a state of dependency from which they have been slow to recover.'[72]

Contemporary forms of neocolonialism are articulated in economic terms but are closely interwoven with, and are also communicated through, consumer culture and cultural capital. The prioritizing of westernized products during colonialism led to a cultural privileging of western goods. The continuous prevalence of western brands over local products in developing countries is a result of a global shift in cultural value – a legacy of imperialism:

> There has been much talk of the 'McDonaldization' or 'Coca-Colaization' of the world as large corporations spread both their production centres and also their sales outlets to more and more remote parts of the globe [...] For some, this spread of 'Western' consumption practices is interpreted as a form of neo-colonialism. 'Non-indigenous' music, food and clothing are promoted as being 'better' and thus those people who can afford such consumer goods are regarded as more 'developed' or 'advanced.'[73]

This alimentary hierarchy is connected to imperial histories, and this food-value economy is expressed in *The Inheritance of Loss*, as will be explored below. Other forms of neocolonialism are also interrogated in Desai's text – for example, economic bodies and forces in the global economy that disproportionately support already developed national (usually western) markets such as nongovernmental organizations (NGOs), the International Monetary Fund (IMF), and the World Bank.[74] These institutions promote 'modernization' and development by granting aid and investment packages, but ultimately aim to financially benefit wealthy investing countries and, more often than not, lead to negative social and environmental consequences for recipient nations. Again, these global politics are often explored through the symbolism of food and eating in Desai's novel. An examination of imperial contexts explains why certain postcolonial nations – in this case, India – suffer from poverty and wealth inequality. Due to India's long and troubled history of famine and food insecurity, there exists an abiding memory of hunger, deprivation, and power inequality that continues to persist in the contemporary moment, and these collective histories structure Desai's narrative.

This historical exposition contributes to the creation of narratives that situate food as an object whose value is determined by both current market assessments and an abiding memory of historical deprivation and oppression. Food becomes a potent sign of power. Nimi's hunger protest responds to this historically constructed meaning of food, focalized through imperial history and the colonial encounter. Her somatic protest is against these legacies – her food refusal is a denouncement of this history and a rejection of the power relations it produces. By subjecting her own body to starvation, she writes the violence and privations of colonial rule onto her body, in a radical affirmation of suffering. Through her own somatic agency, the psychosocial, material, political, and personal suffering produced by colonialism is painfully inscribed on her body.

The Inheritance of Loss: Postcolonial Loss and Hunger

Postcolonial writing often contains images of consumption and extreme hunger.[75] The implications of this hunger have a unique cultural significance within the context of imperialism. Food is culturally and historically encoded as a symbol for national identity and belonging. Colonial legacies are contained within the politics of food and cuisine, and these in turn manifest in *The*

Inheritance of Loss. This novel's narrative presents a history filled with the ramifications and memory of imperialism. The imperial desire for political, economic, and ideological hegemony is reflected in the historical metanarrative of the novel, and this sentiment is also represented in the novel in a variety of neocolonial forms. In *The Inheritance of Loss*, Desai presents a narrative landscape where protagonists are dislocated geographically and ideologically. The characters presented are hybrid – either because of colonial/postcolonial migration or due to imposed/adopted colonial ideologies. These dislocations are articulated and negotiated through food. Representations of hunger, desire, and eating are utilized as the forms through which identities are shown to be in a constant state of establishment and renewal.

Many of the main protagonists of *The Inheritance of Loss* represent a variety of national allegiances and cultural histories, but they do share a common characteristic: a longing to create a distinct cultural identity within the political and cultural context of their geography. This identity is created and reinforced through food and eating practices. The principal characters – Lola, Noni, Uncle Potty, Father Booty, Sai, the judge, the cook, and Biju – populate the novel's global narrative landscape with scenes taking place in England, America, and Kalimpong. The novel stays primarily with this last location, a liminal space filled with complex national identifications. Except for Biju and his father the cook, the characters comprise a small but distinct cultural subset in the multicultural community of Kalimpong. These characters share a commonality that distinguishes them from the rest of the town's inhabitants: they speak in English, and they prefer a western diet that is not only abundant and varied but often consists of imported produce. They go to the local library to check out and consume classic English texts and travel narratives containing romanticized imperial fantasies. Their cultural language is anglicized; however, this anglicization occurs in a manner specific to the hybrid space they occupy.

Although they celebrate Christian festivals such as Christmas and read the literature of the west, the culture they create is a shadow version of the one they attempt to replicate: a hybridized space. "'What's for PUDS?' Lola, when she said this in England, had been unsettled to find that the English didn't understand [...] even Pixie [her daughter residing in England] had pretended to be bewildered."[76] This hybrid space is not the politically productive strategy outlined by Homi Bhabha in *The Location of Culture* – a 'double vision which in disclosing the ambivalence of colonial discourse also disrupts its authority.'[77] The mimicry represented in this text is a denigrated form of imitation, embarrassing and often comedic. These characters' unique culture is an adapted version of the western culture they venerate and seek to emulate, instilling a sense of alienation from their local neighbours. This neither-here-nor-there situation 'condemned them over several generations to have their hearts always in other places, their minds thinking about people elsewhere; they could never be in a single existence at one time.'[78] It is through their cultural preferences that these characters express this hybridized subculture. Their culinary choices and the style in which they eat particularly articulate this identity. As they want to identify with all the associative qualities of western culture – affluence and cultural superiority – these characters opt for westernized foods and cultural habits.

In the novel, food is empowered with a transformative quality: you are what you eat. The consumption of food denotes an assimilation of the value ascribed to the item ingested; it becomes a part of identity. This sentiment is expressed repeatedly within Desai's novel and echoes stadial and transformative theories about food and culture. This assimilative theory of the body is rooted in concepts of both the individuated 'closed' body and the communal 'open' body. In a closed body model, the subject achieves a stable somatic identity through ingesting and then policing the boundaries of the body but is theorized as a relatively static state; once something is ingested, it is retained, and what is not retained is abjected and then ejected into the world outside the body. The open body model, however, considers the somatic locus of the self as more fluid and emergent. The cook expresses this model to Sai in his theories on ethnic characteristics:

> Coastal people eat fish and see how much cleverer they are, Bengalis, Malayans, Tamils. Inland they eat too much grain, and it slows the digestion – especially millet – forms a big heavy ball. The blood goes to the stomach and not to the head.[79]

The cook associates mental acumen with the ingestion of fish, a common belief in South Asia. In this passage, the relationship between the alimentary and character is collapsed. The body is both the location and source of becoming. 'The body's engagement with the world constantly recalibrates it, remaining in balance by the intake and outflow of substances and with whom they interact.'[80] In contrast to this, Uncle Booty makes a case for the diametrically opposed relational body: 'To kill you must be carnivorous or otherwise you're the hunted.'[81] Uncle Booty conveys the possibility of only two subject positions: to eat or to be eaten. This suggests an irreducible relationship between two diametrically opposed subject positions based on difference, and this illustrates theories of more restrictive, individuated body – unsurprising, given Uncle Booty's mimicry of the metropole's culture.

In the construction of consumption or ingestion, food is taken wilfully into the body as the internalization of the meaning associated with the food item. As such it can be interpreted as an ingestion of ideology because food objects and their meanings are socially constructed and reflect the dominant discourses of the geopolitical space from which they emerge. The ideology associated with the material substance digested is performed as an act of acceptance. But contained within this acceptance is a refusal of a myriad of associative qualities of items of food *not* chosen, or *not* eaten. As Bourdieu states:

> Tastes (i.e., manifested preferences) are the practical affirmation of an inevitable difference. It is no accident that, when they have to be justified, they are asserted purely negatively, by the refusal of other tastes. In matters

of taste, more than anywhere else, all determination is negation; and tastes are perhaps first and foremost distastes, disgust provoked by horror or visceral intolerance ('sick making') of the tastes of others.[82]

Contained within the conscious ingestion of food is the disavowal of everything that is understood to discursively oppose it. Taste is not innate or biological. Food is constructed as a sign that is inserted into a dense network of other signs and significations based on difference. In the instance of food, ingestion is as much about desire as it is about disavowal. This means that identities are constructed against other subjects in a process of differentiation. As the identity of the self is shored up in the belief that taste comes from an essential place of personal choice and what is interpreted as 'correct' consumption within the cultural context, so the Other is utilized as a site to dump all that is denied and negated. These bodies are closed – individuated – and rely on a system of difference that mirrors the logic of Othering. The closed body is concerned with what enters the body's borders via the mouth but is also a communicative model that uses food as a semiotic system of difference, structuring the exchange of information and meaning between subjects. In Desai's novel, we can see evidence of both types of bodies, merging and slipping into one another in the colonized space. However, there is an emphasis on the atomized closed body throughout the text – it encroaches into the socio-political time period of colonization – jostling up against the open body – but overwhelmingly prevails when the novel moves into the neocolonial period, defined as it is by the global neoliberal logic of competition, individual consumption, and desire.

The characters in the novel intentionally declare something of their own social, political, and cultural proclivities and identities in the culinary choices they make:

> Food, as a commodity, is consumed not simply for its nourishing or energy-giving properties, or to alleviate hunger pangs, but because of the cultural values that surround it. By the act of purchasing and consuming the food as commodity, those values are transferred to the self. The food is chosen to reflect to oneself and others how individuals perceive themselves, or would like to be perceived. For example, the act of purchasing and eating a Big Mac demonstrates a membership of a cultural group that differs from membership of a cultural group which prefers to dine at an expensive restaurant. That is not to say that the same individual will not engage in both activities, merely to note that the persona thus presented is different in each case. Such uses of commodities are central to the development and articulation of subjectivity. When food is consumed symbolically, its taste is often of relatively little importance: it is the image around the food product that is most important.[83]

Food attributed with a higher level of status and power confers this to the consumer's body. How value is attributed to food is complex and is culturally specific. Food value goes beyond economic value – although this is an important factor as the right to choose the amount, quality, and nature of the food consumed is open only to a privileged class. In the novel, value is attached to certain food items – in this case, westernized cuisine. The higher cultural and economic value attributed to western cuisine can be traced to the mainstay of colonial ideology: that western cultural capital is worth more than 'native' capital. As Bourdieu states in *Distinction*:

> The relations between these classes are ones of competitive striving in which struggles for economic position and for status are connected, as the difference between legitimate tastes and less legitimate ones yields different and unequal stocks of cultural capital for the members of different classes.[84]

Economic rarity and culture intersect to produce a hierarchical system of taste. In a postcolonial context, legacies of imperialism determine a cultural hierarchy that is articulated along the fault lines of native versus the west. This form of neocolonialism leads to the privileging of western cultural capital. The process of eating becomes a demonstrative act of power. This hierarchy is reiterated again and again in Desai's novel: 'cake was better than *laddoos*, fork spoon knife better than hands, sipping the blood of Christ and consuming a wafer of his body was more civilized than garlanding a phallic symbol with marigolds. English was better than hindi.'[85] Even Sai's adolescent love for her tutor Gyan is transformed into prejudice by this hierarchy: 'Eating together they had always felt embarrassed – he, unsettled by her finickiness and her curbed enjoyment, and she, revolted by his energy and his fingers working the dal, his slurps and smacks.'[86] Native modes of consumption produce disgust. Sai's reaction to native eating practices is a means of distinguishing herself from the abject native body.

The polarized hierarchy is upheld by characters from both sides of the global line that constitutes the divide. It is a universal logic. Cultural superiority is communicated through the specificity of culinary preference. The cook colludes in the maintenance of the cultural superiority of western food. '*Angrezi khana* [the cook] [...] was sure since his son was cooking English food, he had a higher position than if he were cooking Indian.'[87] The westerners in the novel follow suit. For instance, the Italian owners of one of the restaurants in which US illegal immigrant Biju works, when searching for employees, hoped 'for men from the poorer parts of Europe – Bulgarians perhaps, or Czechoslovakians. At least they might have something in common with them like religion and skin color, grandfathers who ate cured sausages and looked like them.'[88] In the novel's portrayal of the 'west' – focalized through Biju's experiences as an alien immigrant working in the United States – affluent westerners like to frequent foreign restaurants, but this consumption takes place in the unthreatening Orientalized restaurant space. Although the aesthetics of the restaurant relies on the notion of difference, the consumption of foreign cuisine indicates a mastery of foreignness and cosmopolitanism. The romance of colonial times past is packaged in the apparently hygienic and pristine space of the restaurant. 'Colonial India, free India – the tea was the same, but the romance was gone, and it was best sold on the word of the past.'[89] This stringent dichotomy of power is expressed through the alimentary postcolonial exotic. Food is associated with the easy consumption of culture. All participants in the global food economy uphold its power differentials regardless of whether they are winning.

Expressions of hunger in *The Inheritance of Loss* convey the desire to exert power over places, ideologies, and people. All the characters experience this in some form or another. Sai and Gyan demonstrate the softer side of desire when they give one another food-based pet names. She is his 'momo,' he her 'kishmish'; desiring the absolute union of new lovers, they 'melted into each other like pats of butter.'[90] The besotted tone of the young lovers disguises the desire to assert ownership, while also (at least initially) masking the difference in culture and status their eating habits make apparent: he eats with his hands, she with a knife and fork. The judge experiences a more aggressive desire that manifests as a will to consume a culture, but somewhat ironically, it becomes a desire to consume himself. The moment he leaves Piphit, he becomes exposed to the western ideologies of cultural superiority that invade his mind and turn him against his countrymen and himself.

Jemubhai becomes a foreigner everywhere, including in his own consciousness. The duality he experiences by feeling a 'foreigner in his own country'[91] is expressed as an inward-facing hatred and shame. 'Sometimes, eating that roast bustard, the judge felt the joke might also be on him, and he called for another rum, took a big gulp, and kept eating feeling as if he were eating himself, since he, too, was (was he?) part of the fun.'[92] The judge's identification with Englishness, and his conflicted identification as Indian, manifests in this instance as a self-cannibalization. Lola and Noni express their desire for a more anglicized existence in Kalimpong through their attempts to teach their maid Kesang to cook foreign dishes, and through their importation of British food items. These characters express a desire not only for the basic alimentary pleasure of food consumption but also for mastery of what the food symbolizes.

The consumer/consumed relationship is expressed through a variety of modes in the novel. The last paragraph of the first chapter focuses on national boundaries and deploys a consumptive metaphor, conferring the constitutive embodied logic of eating to the borders of the nation state:

> Here, where India blurred into Bhutan and Sikkim, and the army did pullups and push-ups, maintaining their tanks with khaki paint in case the Chinese grew hungry for more territory than Tibet, it had always been a messy map. The paper sounded resigned. A great amount of warring, betraying, bartering had occurred; between Nepal, England, Tibet, India, Sikkim, Bhutan; Darjeeling stolen from here, Kalimpong plucked from there – despite, ah, despite the mist charging down like a dragon, dissolving, undoing, making ridiculous the drawing of borders.[93]

60 *Kiran Desai's The Inheritance of Loss*

The initial sentiment expressed in this paragraph, at the very start of the novel, foregrounds the vital themes of the book. Hunger is inaugurated as the Will to Power – as a vital and instinctual drive to dominate, achieve, and own. It establishes the complicated historical legacies of colonialism that reinforce the present-day directionality of neocolonialism.

Through plot and a third-person narration, the novel provides a historical grounding and political contextualization that gestures towards a colonial past and how it connects with the present. The narrative makes several far-reaching assertions about the state of the global market that seem as though they are pedagogical asides intended to educate. One is 'The fittest one wins and gets the butter.'[94] And again, rather nebulously:

> The whole system seemed to favor, in fact, the criminal over the righteous. You could behave badly, say you were sorry, you would get extra fun and be reinstated in the same position as the one who had done nothing, who now had both to suffer the crime and the difficulty of forgiving, with no goodies in addition at all.[95]

This is delivered in an unfocalized narrative voice. It is a knowing and generous narrator, who overtly consciously reveals some of the vital themes of the novel. The text targets globalization and its neoliberal logic, and the passage outlines – somewhat comically – the perpetuation of colonial logic in the contemporary moment. 'The first stop was Heathrow and they crawled out at the far end that hadn't been renovated for the new days of globalization but lingered back in the old age of colonization.'[96] Here the start and end of border crossing – the airport – presents a seamless enjambment between past and present.

Possibly the most prominent example of this pedagogic voice is Desai's construction of Biju's America: of the series of restaurants in which he works illegally. Although the aesthetics of these restaurants are meant to communicate a liberal cosmopolitanism, a more unpleasant truth is concealed in the unseen kitchens within:

> Former slaves and natives. Eskimos and Hiroshima people, Amazonian Indians and Chiapas Indians and Chilean Indians and American Indians and Indian Indians. Australian Aborigines [...] [the list of countries continues] [...] Zaireans coming at you screaming colonialism, screaming slavery, screaming mining companies screaming banana companies oil companies screaming CIA spy among the missionaries screaming it was Kissinger who killed their father and why don't you forgive third-world debt; Lumumba, they shouted, and Allende; on the other side, Pinochet, they said, Mobutu; contaminated milk from Nestle, they said; Agent Orange, dirty dealings by Xerox. World Bank, UN, IMF, everything run by white people. Every day in the papers another thing![97]

The narrator descends into this chaotic listing, their accelerating voice echoing the endless and historic conflicts between 'us' and 'them.' The volatility contained in this description motions toward the violence of globalization. It is a force that makes ridiculous the drawing of borders. 'In a period of rapid and rampant globalization and as corporate America finds it more efficient to outsource both capital and labor, the contours of geopolitical nations have been remapped and reconfigured.'[98] The logic of global neoliberalism is summarized astutely by Biju's expat Indian boss Harish-Harry, the owner of the Gandhi Café in New York: 'Find your market. Study your market. Cater to your market. Demand-supply. Indian-American point of agreement. This is why we make good immigrants. Perfect match.'[99] The 'perfect match' that Harish-Harry describes repeats the imperial dynamic between colony and metropole. The text, again, emphasizes the connections between an imperial past and present. Harish-Harry's insistence that Indian immigrants and Americans are perfect matches also reminds the reader that hunger and power operate on a globalized scale, and that native elites are also imbricated in the necolonial, neoliberal system. Food, power, and access are not limited to simply the developed west, but rather older orders of power and economic privilege have been further distributed across the globe, with greater migration and movement across borders perhaps muddying the racial profile of those who have greater food entitlements – nonetheless, the associations between black/brown bodies and endemic hunger, backwardness, and poverty persist.

It is the logic of the market that propels much of the narrative forward in this story. Desai arranges her opposing camps of ideologies and geographies around dividing lines drawn during an age of global colonialism. It is the advent of capitalism, of the global market system, that provides the historical context for both the novel's past and present narrative. The world these characters inhabit is one that is fraught with competition. Resources and power have been allocated largely to the west – and not merely geographically, but ideologically and culturally as well. The consequence of this leads Biju and the judge to migrate. It is what defines Lola and Noni's choice of food and lifestyle, and Uncle Potty and Father Booty's manner of speaking and dress. It is this hierarchy that convinces the cook that Biju will come back fat and successful from the US. This imbalance is kept intact by the logic of the market. The divisive logic of west and Orient, of separate nation states, is consigned to an imperial past – however, imperial legacies persist in the present. These permutations of power are articulated through the language of food, eating, and hunger.

In the final scenes of the book, the economic logic that determines food security and cultural superiority in Kalimpong is inverted by political violence. The GNLF's forcible seizure of power, described in the denouement of the book, causes social and economic disruption in Kalimpong, including a sudden change in the dietary habits of its inhabitants. This signals a transfer of political and cultural power from those who had hitherto possessed it to those who had lacked it. The judge, and the other affluent characters in the novel, experience food insecurity. 'For the first time, they in Cho Oyu [the judge's home] were eating the real food of the hillside.'[100] The judge, Sai, Uncle Potty, Noni, and Lola – the protagonists who in particular expressed their desire for 'elsewhereness' through their dietary choices of imported goods and western-style cuisine – are forced through necessity to turn to the land in order to sustain themselves. While affluent members of Kalimpong's community compete for meagre pickings on the black market, 'the army was still being well fed, and the wife informed her husband that they had been allotted so much butter that they could share it with their extended family.'[101] Aptly, the transfer of political power is expressed as an inversion of food entitlements. Moreover, it is not only political power that is being reconfigured, but also the associative cultural capital attached to colonial habits. While returning from a trip to the library, Sai, Uncle Potty, Father Booty, Noni, and Lola are accosted and subjected to a random search by the GNLF. The insurgents seize their English classics (Trollope and Brontë), and confiscate and callously discard Father Booty's homemade cheese ('"What is that smell? [...] Throw it out [...] It's gone bad."'[102]). Here, the association between the literature of the colonialists and their food is emphasized, and the rejection of both amounts to a rejection of colonial and neocolonial ideology.

This chapter explored the connections between the body, national identity, history, and power. In the case of the postcolonial body, subjectivity is fraught with difficulty and contradiction. The judge attempts to resolve his own fractured self by enacting a vicarious hunger strike upon the body of his wife. The strike is structured upon a relationship between the body and language – the relationship is assimilative, communicative, and mutually constitutive. Nimi protests by refusing both food *and* words – both the material and the semiotic. This chapter, and the close reading of Nimi, has set up many of the vital themes and elements in the examination of the hunger strikes in the following chapters. Although Nimi's hunger strike is brief, it establishes a central and repeated link between the body, identity, power, food, and language that structures the food protests explored in this study.

Notes

1. Melissa Dennihy, 'Globalizations Discontents: Reading "Modernity" from the Shadows', in *Critical Responses to Kiran Desai*, ed. by Sunita Sinha and Bryan Reynolds (New Delhi: Atlantic Publishers, 2004), 1–20 (p. 2).
2. Kiran Desai, *The Inheritance of Loss* (London: Penguin Group, 2006), p. 9.
3. See Mikhael Bakhtin, *Rabelais and His World*, trans. by Hélène Iswolsky (Cambridge, MA: MIT Press, 1968).
4. Pasi Falk, *The Consuming Body* (Thousand Oaks, CA: SAGE Publications, 1994), p. 25.
5. Christine A Hastorf, *The Social Archaeology of Food: Thinking about Eating from Prehistory to the Present* (New York: Cambridge University Press, 2017), p. 282.
6. Falk, p. 25.
7. The apprehension of 'the judge' as concept rather than human is emphasized in Desai's decision to leave the 'j' in his name uncapitalized. The title refers not only to his service in the CIS, but also his characterization as critical, both towards others and himself.
8. Desai, p. 37.
9. Ibid., p. 37.
10. Ibid., p. 38.
11. Ibid., p. 37.

12 Ibid.
13 Ibid., p. 39.
14 Ibid., p. 40.
15 Ibid.
16 Ibid., p. 117.
17 Frederick Cooper, *Colonialism in Question: Theory, Knowledge, History* (Berkeley: University of California Press, 2005), p. 24.
18 Frantz Fanon, *Black Skin, White Masks*, trans. by Charles Lam Markman (London: Pluto Press, 1986), p. 88.
19 Ibid., p. 12.
20 Ibid., p. 12.
21 Desai, p. 38.
22 David Spielman, '"Solid Knowledge" and Contradictions in Kiran Desai's *The Inheritance of Loss*', *Critique: Studies in Contemporary Fiction*, 51.1 (2001), 74–89 (p. 77).
23 Desai, p. 170.
24 Akeia A.F. Benard, 'Colonizing Black Female Bodies Within Patriarchal Capitalism: Feminist and Human Rights Perspectives', *Sexualisation, Media & Society*, 2.4, (2016), 1–11 (p. 3).
25 Desai, p. 171.
26 Suzanne Daly and Ross G. Forman, 'Introduction: Cooking Culture: Situating Food and Drink in the Nineteenth Century', *Victorian Literature and Culture*, 36.2 (2008), 363–373 (p. 262).
27 Beeton, Isabella, *Mrs Beeton's Book of Household Management*, ed. by Nicola Humble (Oxford: Oxford University Press, 2000), p. 363.
28 Deborah Lupton, *Food, the Body, and the Self* (Thousand Oaks, CA: SAGE Publications, 1996), p. 9.
29 Language here refers to everyday discourse (speaking and writing) and ideological Discourses in the Foucauldian sense.
30 Fanon, p. 18.
31 Maud Ellmann, *The Hunger Artists: Starving, Writing and Imprisonment* (London: Virago Press, 1993), p. 33.
32 Martha Sif Karrebæk, Kathleen C. Riley and Jillian R. Cavanaugh, 'Food and Language: Production, Consumption and Circulation of Meaning and Value', *Annual Review of Anthropology*, 47 (2018), 17–32 (p. 18).
33 Parama Roy, *Alimentary Tracts: Appetites, Aversions, and the Postcolonial* (Durham, NC: Duke University Press, 2010), p. 9.
34 Kathleen C. Riley and Jillian R. Cavanaugh, 'Tasty Talk, Expressive Food: An Introduction to the Semiotics of Food-and-Language', *Semiotic Review*, 5 (2017), www.semioticreview.com/ojs/index.php/sr/article/view/1/71 [accessed 20 April 2018].
35 Spielman, p. 77.
36 Desai, p. 172.
37 Ibid., p. 173.
38 Amanda Machin, 'Hunger Power: The Embodied Protest of the Political Hunger Strike', *Interface: A Journal on Social Movements*, 8.1 (2016), 157–180 (p. 157).
39 Lupton, p. 9.
40 Machin, 'Hunger Power', p. 159.
41 David Arnold, *Famine: Social Crisis and Historical Change* (Oxford: Basil Blackwell, 1988), p. x.
42 Ibid., p. ix.
43 Social scientific approaches to food insecurity have a specific goal of hunger alleviation, and thus are necessarily political in tone and purpose. See *Ethics, Hunger and Globalization: In Search of Appropriate Policies*, ed. by Per Pinstrup-Anderson and Peter Sandøe (Dordrecht: Springer Science and Business Media, 2007).

44 Margaret Kelleher, *The Feminization of Famine: Expressions of the Inexpressible?* (Durham, NC: Duke University Press, 1997), p. 5.
45 Arnold, p. 2.
46 Mike Davis, *Late Victorian Holocausts: El Niño Famines and the Making of the Third World* (London and New York: Verso, 2001), p. 6.
47 Davis provides a summary of the exceptional weather pattern responsible for the extended period of drought known as an 'El Niño event'. For further information, see Davis, p. 17.
48 See Thomas Robert Malthus, *An Essay on the Principle of Population* (New York: Prometheus Books, 1998), pp. 2–16.
49 See Michael Turner, *Malthus and His Time* (London: Macmillan Press, 1986), pp. 4–16.
50 Amartya Sen, *Poverty and Famines: An Essay on Entitlements and Deprivation* (Oxford: Clarendon, 1982), p. 1.
51 Bruce Currey, quoted in Stephen Devereux, *Theories of Famine* (New York: Harvester Wheatsheaf, 1993), p. 9.
52 Arnold, p. 7.
53 Deepika Bahri, 'Postcolonial Hungers', in *Food and Literature*, ed. by Gitanjali G. Shahani, Cambridge Critical Concepts (Cambridge: Cambridge University Press, 2018), 335–352 (p. 338).
54 Daniel Maxwell, '21st century famines have nothing to do with a lack of food', World Economic Forum (2017), www.weforum.org/agenda/2017/03/21st-century-famines-have-nothing-to-do-with-a-lack-of-food [accessed 6 October 2018].
55 Parama Roy, *Alimentary Tracts*, p. 124.
56 Adam Smith, *An Inquiry into the Nature and Causes of the Wealth of Nations* (London: T. Nelson and Sons, 1852), p. 184.
57 Malthus, p. 78.
58 Roy, p. 126.
59 Davis, p. 7.
60 Karl Polyani, *The Great Transformation* (Boston, MA: Beacon Press, 1944), p. 160.
61 Davis, p. 9.
62 Ibid., p. 19.
63 Achille Mbembe, 'Necropolitics', *Public Culture*, 15.1 (2003), 11–40 (p. 39).
64 Davis, p. 28.
65 Ibid., p. 26.
66 James Vernon, *Hunger: A Modern History* (Cambridge, MA: Harvard University Press, 2007), p. 49.
67 Davis, p. 16.
68 Vernon, p. 6.
69 See Anne Buchanan, *Food, Poverty and Power* (Nottingham: Spokesman, 1982), pp. 32–43.
70 Amiya Kumar Bagchi, *The Political Economy of Underdevelopment* (Cambridge: Cambridge University Press, 1993), p. 94.
71 India is now designated as a developing or, in some instances, a developed nation, but struggles with a very large wealth gap.
72 Jon Bennett with Susan George, *The Hunger Machine* (Cambridge: Polity Press, 1987), p. 116.
73 Katie Willis, *Theories and Practices of Development* (New York: Routledge, 2005), p. 193.
74 See Joseph Stiglitz, *Globalization and Its Discontents* (London: Penguin Press, 2002), pp. 195–213.
75 A brief note on terminology around hunger: note that the difference between starvation and famine is slight. Starvation is the exacerbation of long-term chronic hunger within a specific geographic area. Famine is a result of unchecked

starvation that leads to a breakdown in social functioning in multiple arenas that is not limited to starvation-related deaths, and generally famines are related to high mortality rates. Both starvation and famine are characterized by extreme hunger. 'Chronic hunger and its attendant malnutrition are "silent" emergencies found among impoverished people globally that seldom attract urgent intervention. Starvation and famine are generally more confined and concentrated than endemic undernutrition, chronic hunger, and malnutrition': John R. Butterly and Jack Shepherd, *Hunger: The Biology and Politics of Starvation* (Lebanon, NH: University Press of New England, 2010), p. 29. Irrespective of the particular scale of the hunger experienced in these definitions, they are situated on the same continuum characterized by a lack of food so insistent that it interferes with the functioning of people and societies, to varying degrees.

76 Desai, p. 152.
77 See Homi Bhabha, *The Location of Culture* (London: Routledge, 1994), p. 88.
78 Desai, p. 311.
79 Ibid., p. 73.
80 Hastorf, p. 282.
81 Desai, p. 195.
82 Pierre Bourdieu, *Distinction: A Social Critique of the Judgement of Taste*, trans. by Richard Nice (Abingdon: Routledge, 2010), p. 49.
83 Lupton, p. 26.
84 Bourdieu, p. xx.
85 Desai, p. 30.
86 Ibid., p. 176.
87 Ibid., p. 17.
88 Ibid., p. 48.
89 Ibid., p. 133.
90 Ibid., p. 129.
91 Ibid., p. 29.
92 Ibid., p. 134.
93 Ibid., p. 9.
94 Ibid., p. 134.
95 Ibid., p. 134.
96 Ibid., p. 134.
97 Ibid., p. 134.
98 Anita Mannur, *Culinary Fictions: Food in South Asian Diasporic Culture* (Philadelphia, PA: Temple University Press, 2010), p. 184.
99 Desai, p. 145.
100 Ibid., p. 281.
101 Ibid., p. 293.
102 Ibid., p. 217.

Bibliography

Arnold, David, *Famine: Social Crisis and Historical Change* (Oxford: Basil Blackwell, 1988)

Bagchi, Amiya Kumar, *The Political Economy of Underdevelopment* (Cambridge: Cambridge University Press, 1993)

Bahri, Deepika, 'Postcolonial Hungers', in *Food and Literature*, ed. by Gitanjali G. Shahani, Cambridge Critical Concepts (Cambridge: Cambridge University Press, 2018), 335–352

Bakhtin, Mikhael, *Rabelais and His World*, trans. by Hélène Iswolsky (Cambridge, MA: MIT Press, 1968)

Beeton, Isabella., *Mrs Beeton's Book of Household Management*, ed. by Nicola Humble (Oxford: Oxford University Press, 2000)

Benard, Akeia A.F., 'Colonizing Black Female Bodies Within Patriarchal Capitalism: Feminist and Human Rights Perspectives', *Sexualisation, Media & Society*, 2. 4 (2016), 1–11

Bennett, Jon with George, Susan, *The Hunger Machine* (Cambridge: Polity Press, 1987)

Bhabha, Homi, *The Location of Culture* (London: Routledge, 1994)

Bourdieu, Pierre, *Distinction: A Social Critique of the Judgement of Taste*, trans. by Richard Nice (Abingdon: Routledge, 2010)

Buchanan, Anne, *Food, Poverty and Power* (Nottingham: Spokesman, 1982)

Butterly, John R. and Shepherd, Jack, *Hunger: The Biology and Politics of Starvation* (Lebanon, NH: University Press of New England, 2010)

Cooper, Frederick, *Colonialism in Question Theory, Knowledge, History* (Berkeley: University of California Press, 2005)

Daly, Suzanne and Forman, Ross G., 'Introduction: Cooking Culture: Situating Food and Drink in the Nineteenth Century', *Victorian Literature and Culture*, 36. 2 (2008), 363–373

Davis, Mike, *Late Victorian Holocausts: El Niño Famines and the Making of the Third World* (London: Verso, 2001)

Dennihy, Melissa, 'Globalizations Discontents: Reading "Modernity" from the Shadows', in *Critical Responses to Kiran Desai*, ed. by Sunita Sinha and Bryan Reynolds (New Delhi: Atlantic Publishers, 2004), 1–20

Desai, Kiran, *The Inheritance of Loss* (London: Penguin Group, 2006)

Devereux, Stephen, *Theories of Famine* (New York: Harvester Wheatsheaf, 1993)

Ellmann, Maud, *The Hunger Artists: Starving, Writing and Imprisonment* (London: Virago Press, 1993)

Falk, Pasi, *The Consuming Body* (Thousand Oaks, CA: SAGE Publications, 1994)

Fanon, Franz, *Black Skin, White Masks*, trans. by Charles Lam Markman (London: Pluto Press, 1986)

Hastorf, Christine A., *The Social Archaeology of Food: Thinking about Eating from Prehistory to the Present* (New York: Cambridge University Press, 2017)

Karrebæk, Martha Sif, Riley, Kathleen C. and Cavanaugh, Jillian R., 'Food and Language: Production, Consumption and Circulation of Meaning and Value', *Annual Review of Anthropology*, 47 (2018) 17–32

Kelleher, Margaret, *The Feminization of Famine: Expressions of the Inexpressible?* (Durham, NC: Duke University Press, 1997)

Lupton, Deborah, *Food, the Body, and the Self* (Thousand Oaks, CA: SAGE Publications, 1996)

Machin, Amanda, 'Hunger Power: The Embodied Protest of the Political Hunger Strike', *Interface: A Journal on Social Movements*, 8. 1 (2016), 157–180

Malthus, Thomas Robert, *An Essay on the Principle of Population* (New York: Prometheus Books, 1998)

Mannur, Anita, *Culinary Fictions: Food in South Asian Diasporic Culture* (Philadelphia, PA: Temple University Press, 2010)

Maxwell, Daniel, '21st century famines have nothing to do with a lack of food', World Economic Forum (2017), www.weforum.org/agenda/2017/03/21st-century-famines-have-nothing-to-do-with-a-lack-of-food [accessed 6 October 2018]

Mbembe, Achille, 'Necropolitics', *Public Culture*, 15. 1 (2003), 11–40

Pinstrup-Anderson, Per and Sandøe, Peter, eds., *Ethics, Hunger and Globalization: In Search of Appropriate Policies* (Dordrecht: Springer Science and Business Media, 2007)

Polyani, Karl, *The Great Transformation* (Boston, MA: Beacon Press, 1944)
Riley, Kathleen C. and Cavanaugh, Jillian R., 'Tasty Talk, Expressive Food: An Introduction to the Semiotics of Food-and-Language', *Semiotic Review*, 5 (2017), www.semioticreview.com/ojs/index.php/sr/article/view/1/71 [accessed 20 April 2018]
Roy, Parama, *Alimentary Tracts: Appetites, Aversions, and the Postcolonial* (Durham, NC: Duke University Press, 2010)
Sen, Amartya, *Poverty and Famines: An Essay on Entitlements and Deprivation* (Oxford: Clarendon, 1982)
Smith, Adam, *An Inquiry into the Nature and Causes of the Wealth of Nations* (London: T. Nelson and Sons, 1852)
Spielman, David, '"Solid Knowledge" and Contradictions in Kiran Desai's The Inheritance of Loss', *Critique: Studies in Contemporary Fiction*, 51. 1 (2001), 74–89
Stiglitz, Joseph, *Globalization and Its Discontents* (London: Penguin Press, 2002)
Turner, Michael, *Malthus and His Time* (London: Macmillan Press, 1986)
Vernon, James, *Hunger: A Modern History* (Cambridge, MA: Harvard University Press, 2007)
Willis, Katie, *Theories and Practices of Development* (New York: Routledge, 2005)

3 The Text, Starving Body, and J.M. Coetzee's *Life & Times of Michael K*

South African author J.M. Coetzee has enjoyed an immense amount of both critical and popular success as a contemporary author. His work is highly considered in scholarly circles and is taught regularly in schools and universities as part of the postcolonial canon. Coetzee's literary fiction lends itself to critical and philosophical readings; his typically opaque narratives encourage interpretive and allegorical engagement, and they can also be productively read through postcolonial theory. This chapter focuses on his Booker Prize–winning novel *Life & Times of Michael K*, published in 1983. The novel is set in an imagined South African socio-political landscape, and its sparse narrative is a portrayal of an equally meagre character: Michael K. Throughout the novel, Michael K displays various atypical behaviours that involve a rejection of community, speech, and food. These anomalous behaviours are a response to demands placed on the character to define himself and his place within the civil unrest that serves as the novel's setting. These demands operate within the plot, and in the extratextual processes of reading the narrative itself. Both kinds of demands are mediated through a discourse of power that is focalized through a central motif in the novel – hunger.

I read hunger as a biotechnology that mediates Knowledge/Power as it is situated in the novel's narrative, and as an expression of socio-political discipline upon the politicized canvas of the black/coloured South African body. As Elleke Boehmer reminds us, '[i]n colonial representation, exclusion or suppression can often literally be seen as "embodied."'[1] Hunger and eating are functions of a wider discourse of biopolitics found in the novel. Hunger is expressed as a desire to know, discipline, and exert power. Within *Life & Times of Michael K*, various individuals and socio-political institutions 'hunger' after the novel's protagonist, Michael K. The imagery of consumption is deployed in the text as a desire to render Michael's obvious physical and subjective alterity – his Otherness – intelligible. Michael K attempts to evade these appetites using various physical and linguistic strategies, most demonstrably his hunger strike, which is the focus of much of this chapter. Through this act of apparent embodied agency, the novel can be read as an exploration of responses – both discursive and embodied – to fractured and subjugated colonial subjectivity, and to fractious national and racial identity in the divided nation state. Hunger, and its associated practices, is

DOI: -3

interpreted as simultaneously productive and destructive, signifying as a somatic process that contains the central contradiction of the native body and its place within the colonial nation state. The power that hunger allegorizes asserts its influence on Michael K, on the various figures he encounters in the text, and on the readers of the novel itself.

I explore the dynamics between representation and materiality (as analogous to discourse or mind/body) through the specific forms and outcomes of Michael K's bodily protest. Michael K figures the material constraints of his body, and the oppression of the colonial state, as a prison to be escaped. Like all the hunger strikers examined in this book, Michael's self-imposed hunger is constructed in response to the contradictions and dislocations of the colonial/native body, as they are read through the binaries of the Cartesian dualism. Michael K attempts to transcend the site of these contradictions – his body – by subjecting it to a process of self-starvation, and by removing it from various spaces and gazes that attempt to interpret and thus possess it. The hunger strike can be read most productively as a site through which we may understand the valences of power in the colonial state that contains both political agency and limitations. This chapter reads the exclusionary practices in the novel – self-imposed by Michael K and those exerted upon him by others – as vital for understanding the politics of the novel and the South African context allegorized within it.

Ahistoricism and Allegory

Despite his critical success, the reception of J.M. Coetzee's literature has not been without controversy. The accusations[2] levelled at his work often denounce its muted and anomalous protagonists, inconclusive narratives, and imagined geographical and historical backdrops as specifically apolitical – an accusation that is particularly damning within the socio-political context from which Coetzee writes. South African authors are often expected to bear the ethical burden of representing South Africa in politically productive ways. A review from the pages of *African Communist* lambasted *Life & Times of Michael K*, claiming:

> The absence of any meaningful relationship between Michael K and anybody else [...] means that in fact we are dealing not with a human spirit but an amoeba, from whose life we can draw neither example nor warning because it is too far removed from the norm, unnatural, almost inhuman. Certainly those interested in transforming South African society can learn little from the Life & Times of Michael K.[3]

The author of this review objects to Coetzee's oblique allusions to the politics of colonial resistance and his failure to clearly enunciate anti-colonial strategies. Coetzee's own insistence on his art's autonomy from the political only deepens this accusation of apoliticism and ahistoricism. He claims that '[s]torytelling is another, an other mode of thinking' and should not become simply a 'message with a rhetorical or aesthetic covering.'[4] More recent critical works have attempted to counter this position. In his article 'The (Im)possibility of Ecocriticism,' Dominic Head claims that *Life and Times of Michael K* 'reorients the politics of post-colonialism from an ecological perspective'[5] and so the novel presents useful theoretical intersections between postcolonial studies and ecocriticism.[6] However, Anthony Vital counters this reading of the novel in his article 'Toward an African Ecocriticism':

> It is attention to how the narrative form so persistently develops K as solitary, as isolate, that signals K as an implausible carrier of ecological value [...] [as] crucial to ecological discourse is the idea of relation: ecology's scientific work is predicated on the recognition that living beings exist only in relation to other living beings as well as to a complex nonliving material order.[7]

Vital claims that to comment on the relationships between humans and their ecological surroundings, Michael K must be inserted within a network of others — be in and interact with a community — something Michael scrupulously avoids in the novel. Compounding Michael K's avoidance of society, the narrative of the novel itself is seen as removed from the social and historical world in which it exists — a situation created by the particular distancing techniques employed by many of Coetzee's narratives, as indicated by Derek Attridge:

> Their distance — with the exception of *Age of Iron* and *Disgrace* — from the time and place in which they were written, the often enigmatic characters (the barbarian girl, Michael K, Friday, Vercueil, and many others), the scrupulous avoidance of any sense of an authorial presence, the frequently exiguous plots: all these encourage the reader to look for meanings beyond the literal, in a realm of significance which the novel may be said to imply without ever directly naming.[8]

Attridge's quote gestures to a gap between text and object, or text and the world. The accusation that Coetzee's texts fail to produce productive modes of political thought has momentum in a range of contemporary critical discussions of his work — past and present. This chapter considers both arguments. Although *Life & Times of Michael K* may seem dislocated and 'out of history,' I argue this does not render the text apolitical. In fact, the novel is rooted in the real, although its grounding in and representation of the 'real' is by no means straightforward, and this does somewhat muddy the novel's political messages and potential.

The novel must be situated within a well-established debate about Coetzee's work that is concerned with scrutinizing the relationship between textual representation and the material world it signifies. *Life and Times of Michael K* is preoccupied with this binary, which I read as being reflected metaphorically in the Cartesian duality that explicitly structures Michael K's hunger strike. The debate around materiality and signification has defined the parameters of much critical work on *Life & Times of Michael K*. As David Atwell posits in his collected thoughts on Coetzee's oeuvre, *South Africa and the Politics of Writing*:

> Since [Teresa] Dovey's study, a number of essays have taken up her implicit call to postmodernist and poststructuralist theory, with the consequence that we now have a considerably oversimplified polarization between, on the one hand, those registering the claims of political resistance and historical representation (who argue that Coetzee has little to offer) and, on the other, those responsive to postmodernism and poststructuralism to whom Coetzee, most notable in *Foe*, seems to have much to provide.[9]

The views expressed by Vital and *African Communist* are based on the fact that Michael K is an anomaly: isolated from the world, he cannot provide a productive model of anti-colonial struggle. The mode of Michael K's existence, articulated as a constant avoidance of all forms of society, precludes the novel's potential for political interrogation. However, according to Atwell, taking up a bipartisan stance on his work also leads to a problematic critical juncture characterized by an oversimplified binary – the material real versus the discursive real.

Critics who claim that Coetzee's refusal to temporally and geopolitically situate his work amounts to political negligence often situate his writing within a racialized hierarchy of taste and literary reception. Coetzee's writing, says Vaughan, 'can say next to nothing, and certainly nothing reliable, about experiences outside the modality of its own racial-historical dialectic.'[10] This assessment reduces Coetzee's texts to an insular mode of white South African writing that is only preoccupied with an intellectual and elite audience. His work is placed in a context of aesthetic autonomy that situates literature outside of the world of politics, such as those presented in South Africa. This kind of writing, according to Michael Chapman, 'confirm[s] the suspicions of many black writers that literary pursuit in white South Africa has rather more to do with the gratifications of libidinal language than the fulfillments of fighting political injustice.'[11] Coetzee is read as an author whose literary sensibilities affirm European literary traditions and deliberately seek out a privileged audience. As Alys Moody states:

> Addressing himself as a provincial writer to the metropolitan centers of power [...] [Coetzee] seeks to locate himself within literary systems of value that speak above all to the value of autonomy, and to find a global audience that prizes this concept where his local peers do not.[12]

In this light, *Life and Times of Michael K* presents itself as a difficult source of political resistance. Its allegorical narrative is outward-facing and seemingly uninterested in local contexts and problems. Thus, his work is viewed through a lens of political reluctance and a retrograde commitment to European literary traditions. According to these readings, his writing seeks to transcend its own geopolitics, rather than find useful literary significations within its own matrix.

So how can we align an ahistorical and politically hesitant *Life and Times of Michael K* and its isolate protagonist with the radical aims of the hunger strike? Critics have often used the lack of realism in Coetzee's narratives as proof of his lack of political responsibility. The overly allegorical mode in which he writes, they argue, widens the gap between literature and its engagement with the world. This kind of postmodern writing operates discursively above the referents to the real, and from such writing an anti-colonial politics is impossible to extract. '[A] vant-garde literature was considered "complicit or, at best, politically irrelevant."'[13] A possible response to the opposition implied here, between realism and allegory, between what is real and what is *too* discursive (amounting only to textual play and abstraction) can be negotiated using a poststructural framework. As Louise Bethlehem states in response to Chapman's denouncement of Coetzee's writing ('masturbatory release' is the phrase used by Chapman): 'Riding on the implicit hierarchization of values inherent in the libido-versus-lived-reality opposition, Chapman's criticism openly sides with the "real."'[14] If South African writing's claim to political urgency is tied to realist modes of writing, then the material is privileged over language, which – according to Bethlehem – can be quickly deconstructed by acknowledging realism's mimetic limits. Besides this, the sorts of accusations levelled at Cotezee downplay the European influence on the form of realist narratives. 'This doctrine leads in a straight line of descent from George Eliot to South African appropriations of realism.'[15] Notwithstanding realism's cultural heritage, the form also has the tendency to fix, rather than destabilize, historical teleology and emphasizes an ability to represent its wholeness. As such, it can be read as a curious form with which to create an anti-colonial or postcolonial aesthetic, which is committed to alternative visions of history and knowledge. '[T]he confident retrieval of History as referent obscures the extent to which the master narratives, the master's narratives, liberalism and realism, are already in crisis with respect to the social, even as they enact their ascendancy.'[16] The claim of authenticity implied by the realist mode is critically interrogated for its ability to reinforce hegemony – historically and formalistically – even in its direct engagement with the political. Far from being more appropriate to represent the politics of the world, realism can undermine rather than reinforce the radicalism required to articulate alternative political futures. Unlike realism, allegory is a more self-aware form of representation that acknowledges its own artifice and use of language, although it is sometimes viewed as a more restrictive form than realism.

In order to revitalize allegory with the necessary political will to situate the politics of hunger within *Life and Times of Michael K*, I turn in the first instance to Stephen Slemon's work on postcolonial praxis, particularly with regard to his comments on the allegorical. Slemon notes:

> [T]he assumption of 'natural' seamlessness within language has never taken hold within colonized territory; for when colonialism transports a language, or imposes it upon a differential world, a fracturing, indeterminate

semantics becomes the necessary medium for verbal and written practice [...] Post-colonial cultures have a long history of working towards 'realism' within an awareness of referential slippage.[17]

Allegory is particularly significant for postcolonial writers because it disrupts the orthodoxy of history as it is reflected in realist modes of writing and colonial representations in general. Distancing his argument from Aijaz Ahmad's[18] retort to Frederic Jameson's claim that all 'third world' literature is necessarily allegory,[19] Slemon suggests that we might rather see allegory as a function of the 'conditions of postcoloniality,' insisting on allegory's productive mode as a re-engineering of dominant western literary forms to convey the fluidity and instability of language and meaning – an appropriate adaptation for the colonial condition. Although Coetzee writes predominantly from and to a European discursive stance, he culturally locates himself within the hybridities of the colonial and postcolonial state; 'a people no longer European, not yet African.'[20] As such, his allegorical writing – although indebted to prevailing European literary traditions – is suited to the representation of the postcolonial condition.

Embracing allegory as an arena for radical expression and political agency is not completely unproblematic, however. Critics of the allegorical mode accuse it of rigidity and claim its forms encourage rationalistic thinking – like solving a puzzle or breaking a code, the allegory's nebulous narrative can appear to manipulate the reader into empiricist thinking, leading to a singular 'correct' answer. But unlike the scientific method, allegory also contains the ability to refract into multiple meanings:

> Thus the negative critical attitude to allegory focuses on its rationalistic proclivity and its authoritarian control of the play of meaning, which implies dryness and rigidity. Yet the same intellectual bent of allegory can be construed in radically different terms as textual openness requiring the reader's active participation.[21]

The place of allegory within a colonial and postcolonial context is an ambivalent one. However, Judith Anderson warns not to position the two modes discussed here – realism and allegory – against one another. 'When mimesis becomes naively conceived realism, or photocopy [...] allegory must correlatively become naively conceived as abstraction. Mind separates from matter, psyche from flesh, concept from history.'[22] To arrange mimesis into two opposing camps reinforces rather than interrogates the epistemologies of literature. Again, as explored above, the privileging of realism, as literary criticism has done since the nineteenth century, privileges a historical metanarrative, as well as a notion of 'the real' that it attempts to mimetically capture. Opposed to the privileged realist narrative, allegory may appear too limiting, prescriptive, and nebulous to say anything 'real.' But by reminding us of the openness of allegorical mimesis, and acknowledging the allegorical nature of all writing (including realism), Anderson argues that we can transcend such binaries (including the Cartesian binaries she mentions above), for 'a denial of allegorical thinking and allegorical form allows binaries to reign.'[23] Moody reminds us of Coetzee's assertion for the aesthetic autonomy of literature as 'another, another mode of thinking,' 'he maintains that such a mode is valuable – and politically powerful – precisely because of its capacity to usurp history's discursive position.'[24] Coetzee's assertion may open up space in readings of *Life and Times of Michael K* and presents an opportunity to interrogate the colonizer's historical metanarrative.

To be clear, this chapter does interpret Michael K's body through the politics and history of South Africa, and it reads the starving body as synecdoche for South African history and its Othered black and coloured subjects, but I do not wish to replicate Frederick Jameson's notion of 'national allegory.' I prefer Slemon's more accommodative definition of allegory, with its fissures and slippages. This chapter presents only one reading of Coetzee's hunger novel, not *the* reading of it. 'Like the post-structuralist text, food is endlessly interpretable, as gift, threat, poison, recompense, barter, seduction, solidarity, suffocation.'[25] My reading tries to unravel and make plain the dialectics of colonial discourse within the narrative; whether or not the author intended to replicate its power structures, or whether his aim was to undermine them and fail, is not its main concern. Although the novel is seemingly geographically located in South Africa and specific areas in the Western Cape are referenced, there is a lack of historical anchors or other non-fictional referents. This creates a sense of an imagined South Africa, with many of the same social problems, but its nebulous setting does not declare itself in any definite sense. This only adds to accusations that Coetzee's writing distances itself from that which is needed to be said, written, and done in the context of anti-colonial and postcolonial politics. However, although the allegorical setting of this book can be read as creating distance between the world and the text, it also accommodates a greater number of textual interpretations – empowering the reader to read with agency. This concept of allegory is prism-like – fracturing instead of forcing narrative toward a singular meaning. In the case of this novel, exact meaning may be harder to pin down, but the novel's form certainly does not preclude political interrogations of the South African state and its colonial history. And certainly, as Moody points out, '[i]n 1980s South Africa, where hunger is a widespread and discursively important effect of the apartheid regime, representations of hunger are politically coded from the outset.'[26] I am interested in these representations and how they can be read in varied ways, characterized by ambivalence. These may be read as a means of reinforcing racist tropes surrounding black bodies and hunger, but Michael K's deployment of the politics of hunger may be read as intentional subversion of these narratives as well. My reading intentionally positions the novel within the geopolitical context of Apartheid and the South African civil war but also examines the novel's preoccupation with representation and its own insular aesthetics. This chapter argues that the novel elicits a pursuit of meaning within the narrative as articulated by hunger and confronts the reader with this hunger for meaning, which itself creates a valuable politics of extratextual self-interrogation.

Histories of Hunger

The history of hunger in South Africa is imbricated with its colonial history, the reverberations of which were felt through the Apartheid era and continue into the present moment. South Africa experienced three official famines between 1912 and 1946, and hunger had become an acute pressing national issue. A few years before Apartheid, colonial authorities established the Native Affairs Department (NAD) to address the growing concern of food security within the country, among other issues concerning the country's black population.[27] With urgent concerns that the deprivation of the labour capacity of the black population would have detrimental effects on the nation's productivity as a whole, along with previous state failures to adequately address previous famines, the NAD was tasked with alleviating state hunger. However, the NAD was generally ineffective in this aim. It sought to rationalize the distribution of funds according to the logic of the free market, which led to ineffective food distribution in light of poor food entitlements. That, combined with corrupt government practices, meant that the NAD proved unsuccessful in addressing hunger – but was able to present as a philanthropic organization and reap the political benefits of this image.

The Apartheid government fared no better. It was estimated that one-third of all non-white children suffered from malnutrition and that, in 1975, between 15,000 and 27,000 children died from starvation.[28] This was also in the context of a country that regularly exported food. This makes clear that food entitlement was the main issue, rather than a lack of food. This lack of entitlement was driven by a racialized inequality, an inheritance from the colonial period and the Apartheid government's stringent application of a cultural racism that intersected with a paternalistic free-market attitude.[29] Charity was interpreted by the state as a deterrent to engendering black and coloured self-sufficiency, and policy reflected these cultural attitudes. Today, food equality is still a pressing issue in South Africa. In 2004, a Human Sciences Research Council report indicated that about 14 per cent of the population is estimated to be vulnerable to food insecurity, and an estimated 25 per cent of children under the age of six have had their development stunted by malnutrition.[30] Although acute instances of food insecurity have been abolished, the hunger that is present is class-based, endemic, and historically connected to previous forms of racist policies enacted by colonial and Apartheid governments.

This food insecurity must be read within a context of racism whereby the black/coloured body is mediated by culturally inscribed tropes. 'Many whites saw African hunger as evidence of basic cultural incompetence. If black people resisted the lessons of modern science and failed to manage their land successfully or to eat intelligently, they were making themselves outcasts from modernity.'[31] Cultural racism – the descendant of scientific racism whose ascendancy had been curtailed by its associations with Nazi racial theory – framed South African hunger in familiar ways. Previous forms of colonial discourse around civilizing the native who did not seem to be able to eat (or exist) properly were deployed in a paternalistic register, casting black and coloured South Africans as requiring education and direction toward Modernity. Simultaneously, a cultural racism that essentialized the black body as inferior and incapable of truly progressing to full Modernity rationalized hunger as a natural outcome of black indolence, the primitivism of their culture, and an inability to innovate and improve. Thus, these two registers of racism operated simultaneously to maintain the divisive logic of the colonial and then Apartheid state. The black/coloured body became an overdetermined site of meaning, historically presupposed as a figure of pity and derision, of needing correction but ultimately requiring stringent administration because it was clear, to the ruling white power, that they would never free themselves from the limits of their own savagery.

Hunger is a vital historical context for situating *Life and Times of Michael K*. The instances of hunger and starvation described in the book have clear historical referents – ones that would have been easily registered by readers of the time. The book does not politicize hunger, because hunger is already politicized from the outset. The hunger represented by Michael K's body simultaneously contains and rejects the colonial tropes of hunger. His vagrant status, his lifelong experience of hunger, his inability to feed himself even when presented with food – these instances can be read as reinforcing the stereotypes of the incompetent native. However, Michael K's politics of food refusal might also be read as a rejection of these stereotypes, or at least an effort to articulate a different narrative with/on the body. If black and coloured South Africans were hungry despite the free market giving them the opportunity to fill themselves, one of the meanings contained within Michael K's starving body is a recuperated history of the failure of the rational market. His constantly open mouth becomes a hole that the rationality of colonial logic vanishes into, exposing its illegitimacy. His starving body also signifies a radical politics of refusal that subverts and operationalizes the politics of sympathy for the wretched of the earth, turning it on its head. Michael's hunger strike is not without its contradictions and ambivalences, which will be explored in greater detail in this chapter's section on the strike itself.

Narrative, Hunger, and the **Life and Times of Michael K**

This section considers how the motif of hunger permeates *Life & Times of Michael K* and the meaning of the various forms it takes. Hunger is constructed in the book as a metaphor of disciplinary biopower, exerted upon Michael K by other characters, socio-political forces, and institutions, and by the reading practices engendered by the form of the novel itself. Eating (and the refusal to eat) is constructed as a biotechnology through which Knowledge/Power is mediated, and this plays out on Michael K's body, most obviously through his self-inflicted hunger. As well as its literal meaning as desire for food, this thematic 'hunger' is demonstrably recognized as a desire for mastery, and as the will to consume in order to internalize, know, and make sense. This can also be restated as a desire for mastery and power. This section explores how the narrative itself is populated by various hungers for Michael K, expressed as a desire to control and mediate his body in the various socio-political spaces depicted in the novel.

The eponymous protagonist of *Life & Times of Michael K* is a difficult man to understand. Michael K's motivations, his self-perceived role in the imagined South African civil war that is the novel's setting, and his feelings or thoughts on the difficulties he faces arising from this setting – essentially, the character's interiority – are obfuscated in Coetzee's text. Over the course of the novel, Michael K shifts from a stable sense of self – 'he had been brought into the world to look after his mother'[32] – into a kind of anti-subject, a slippery representation of undefinable subjectivity, who avoids interpellation or definition. This state of non-being is what renders his subjectivity such an interesting and productive point of inquiry; as the medical officer says, Michael K is 'the obscurest of the obscure, so obscure as to be a prodigy.'[33] This section examines the puzzle that is represented by Michael K and the motivating drive that fuels the desire to solve it: hunger. Somatic necessities and alimentary longings provide the impetus for much of the plot's movement. But hunger is also allegorized as another type of impulse: a particular type of Will to Knowledge that is expressed as a desire to ascribe Michael K's subjectivity a clear definition. This desire, or hunger, is expressed by characters in the novel – they verbally interrogate him, physically control him, and subject him to various forms of surveillance. In other words, they desire to metaphorically consume Michael and digest him as a knowable and contained quantity.

There are several holes and lacunae in *Life and Times of Michael K*. These narrative gaps, situated within the context of escalating civil unrest, present as interruptions to the daily flow of normal life: as suspensions and periods of waiting. These static periods of inaction produce confusion and anticipation in the reader, and introduce the idea of escape contained within the novel – the reader wants to escape the peculiar pacing of the plot, and Michael K attempts his first escape from the city. These periods of narrative stasis, interjected by moments of confusion, stimulate an expectation of narrative logic. Our desire to escape the disquieting narrative form is mirrored in the characters' physical movements across the novel's landscape. The novel opens by focusing on the relationship between Michael K and his ailing mother; her failing health provides the incentive for Michael K's quest to move out of the city. Anna K (and consequently Michael K) are so poor that the social unrest depicted in the novel has already rendered them homeless. Due to the violence and suspension of everyday life in the city, stocks of food become even scarcer, and Michael K's desire to flee is compounded by pressing obligations towards his mother, Anna K. 'How long could he push her around the streets in a wheelbarrow begging for food?'[34] As Michael K and his mother attempt to escape the city for the promise of a better life in the country at Anna K's childhood home, their attempts are constantly thwarted by the flagging bureaucracy of the state, which refuses to provide the necessary permits to facilitate their movement out of the city: 'Everything was suspended while they waited for the permits.'[35] Faced with the 'sea of hungry mouths'[36] that threaten to overwhelm them both if they remain in the city, Michael K and his mother's captive status in Cape Town is characterized by the narrative's primary driving force and the terms through which it is expressed:

hunger. Michael K constructs a romantic ideal of the countryside as the remedy to his and his mother's woes, both psychological and physical, that have been caused by their worsened circumstances in the city. The stagnancy of their situation is manifested via their appetites, both literal and symbolic, for a place where food is abundant, and the fantasy of fulfilment can be realized.

When Michael is being searched at the checkpoint while attempting to move across South Africa, he is found to be lacking the correct documentation and is said to have 'baulked, like a beast at the shambles.'[37] The narrative describes Michael's fear of being 'devoured by time in the camps,'[38] and he moves 'through the intestines of the war.'[39] The terminology that constitutes Michael K's narrative path through the novel is repeatedly presented as alimentary. Individuals and state institutions hunger after Michael K; attempts to objectify him into a functional (or at least quantifiable) subject are in response to the alterity he represents. They want or need to foreclose this alterity, and for Michael K to fit neatly within the narrative of the war. The enigmatic nature of his motivations, his silence in response to this form of control, his incomprehensible refusal to eat in the latter stages of the novel – all of this 'irrational' behaviour encourages the hyper-policing that Michael K experiences in the text. As readers, we collude in this desire – this hunger – as it allegorizes our own desire to understand that which is deemed illogical – that is, that which refuses to obey the hegemonic power of the state, and of the realist novel form.

As hunger is foregrounded as the primary manifestation of desire and control, so the body is located as the site of its significations. The cipher-like Michael K represents the Othered colonial body. He *is* the body in the body/mind divide, in that he is conceived of as primitive – a being concerned only with the baser instincts of the body and its requirements. Per Boehmer:

> The seductive and/or repulsive qualities of the wild or Other, and the punishment of the same, are figured on the body, and as body. To rehearse some of the well-known binary tropes of postcolonial discourse, opposed to the colonizer (white man, West, center of intellection, of control), the Other is cast as corporeal, carnal, untamed, instinctual, raw and therefore also open to mastery, available for use, for husbandry, for numbering, branding, cataloguing, description or possession.[40]

Colonial discourse assigns savagery to the colonial body to maintain the logic of the colonial civilizing discourse, so that intellect and rationality can be rightfully attributed to the colonial masters. Michael K's embodied self signifies a traumatic colonial and postcolonial history that is silenced (corresponding to the cipher-like 'nought' of a figure Michael K is in the novel) within both the novel's narrative and the historical grand narrative of colonialism in which it is situated. The black body is overdetermined into silence, and so cannot speak beyond the body/mind dualisms of colonial discourse. Anna K's body is similarly signified. Her body has become deformed due to an extended sickness. Her distended body implicates Michael K's body and recruits him into a relationship of dependency whereby the needs of her body take precedence over his. Described as parasitic by the medical officer in Chapter Two of the novel ('I also think of her sitting on your shoulders eating out your brains'[41]), it is the needs of Anna K's body, and his – made more urgent in an atmosphere of conflict and food insecurity – that forges the bond between them. The deterioration of Anna's body leads to greater demands on Michael K and draws him closer to her body's visceral needs. 'The needs of her body became a source of torment.'[42] The lack of subject specificity in the syntax indicates an ambiguous receiver of this torment; Michael K and Anna K (and perhaps even the reader) each share in the discomfort relayed in the abject descriptions of Anna K's body.

Michael K is wary of the pressure his mother's bodily needs place upon him. At the start of the novel, before he and Anna start their journey out of the city, Michael K notes that he does 'not like the physical intimacy that the long evenings in the tiny room forced upon the two of them […] he found the sight of his mother's swollen legs disturbing.'[43] However, he does not shirk his duties as a son; after all, he knew 'he had been brought into the world to look after his mother.'[44] Michael K's subjectivity is defined by Anna K, the demands of her body transferring the onus of survival on to his. He is, in a sense, filled with another's subjectivity. In these initial stages of the novel, Michael K only comes into being and is read through his connection with Anna K.

> Just as he had believed all the years in Huis Norenius [the orphanage that Michael K grew up in] that his mother had left him there for a reason which, if at first dark, would in the end become clear, so now he accepted without question the wisdom of her plan for them.[45]

Before his mother's death, Michael K accepts the teleological nature of narratives – his own, and through this the teleology of the novel and history itself. His mother secures him to this belief, and he has very little agency over it. He is a body already inserted within a familiar narrative of bodies – he is not one to question or debate. He withstands his distaste for her body to continue to perform the only identity he has ever known. Anna K's bloated and fat body stands in for the oppressive hungers that Michael K fears, and that he attempts to evade in later stages of the novel. Her hunger for Michael K mirrors other characters' hunger to co-opt, control, demarcate, fully understand, and penetrate his subjectivity – particularly so for the medical officer. He too hungers for Michael – hungers for explanations and an understanding of him. This hunger foreshadows Michael K's food abnegation and provides a rationale for his evasion of society.

Michael K's movement across the narrative demonstrates his desire to escape this coercive hunger. As he embarks on his first episode of food abnegation, the narrator describes the sense of freedom he experiences. He eats rarely and experiences 'spells of airiness […] he felt weaker than before but not sick.'[46] It is during this time, free from the responsibilities to his mother's body, and in isolation, that Michael K experiences a sort of freedom, arising partly from his rejection of the needs of his own body and a denial of hunger – his own, as well as the hungers of others. But as soon as he comes back into contact with other people, his hunger – or desire – is felt once more. 'As he came closer the smell of frying bacon made his stomach churn.'[47] Similarly, upon being captured and instituted into the Jakkalsdrif work camp, Michael K's desire for food re-emerges after a long period of inactivity – again, only when he is reinserted into prescribed spaces of society, among other individuals:

> An orderly came in with a trolley. Everyone got a tray except K. Smelling the food, he felt the saliva seep into his mouth. It was the first hunger he had known in a long time. He was not sure that he wanted to become a servant to hunger again; but a hospital it seemed, was a place for bodies, where bodies asserted their rights.[48]

This excerpt acknowledges a conscious splitting between Michael K's identity and his body (which will be discussed in detail in the next section of this chapter), and it illustrates the reassertion of a contingent body and identity: Michael K only seems troubled by his own somatic needs when in the presence of others. These examples indicate that the 'closed' body – the body that is constituted in negative relation to other bodies – is the source of Michael K's suffering, and his somatic borders are jeopardized particularly when food is involved. Consumption exposes itself as the process, and food as the agent, of union and differentiation. It acts deconstructively, simultaneously revealing the inter-reliance of subjectivity during the process of Othering while providing the means through which distinction is achieved. It is an emergent and unstable process. In this particular context, food is the binding agent that allegorizes Michael K's relationship to others, in a manner he wishes to avoid. Michael K yearns to be free of the anxieties produced by the invasions and over-determinations of his body. Consequently, it logically follows that food abnegation should be Michael K's weapon of choice in attempting to escape them.

In *The Hunger Artists*, Maud Ellmann explores the links between knowing and eating: 'Kierkegaard is only one of many thinkers who implicate digestion in cognition, for the analogy between these processes is integral to Western thought.'[49] Tracing linkages between eating and knowing by critical thinkers ranging from Marx to Freud, Ellmann traces a genealogy of hunger. Ingestion is figured as the truest path to knowledge; to eat something is to literally and figuratively internalize it, to fully assimilate its essential properties. Knowledge, conceived as nourishing object, is posited in *Life & Times of Michael K* as the aim of characters who attempt to consume – and so know – Michael K. The text's deployment of eating metaphors clusters around two forms of interaction with its protagonist. The first is the attempted confinement of Michael K in various disciplinary ways, and in controlled spaces in an effort to create a Foucauldian docile body: 'a docile body that may be subjected, used, transformed and improved.'[50] This includes Huis Norenius, the orphanage that Michael grows up in; the hospital; the work gang; Visagie farm; Jakkalsdrif; and Kenilworth. These spaces usually employ some form of repetitive physical activity and the control of space and time – highly effective means, according to Foucault, of exerting the type of biopower that creates and mediates a politically desirable passive subject. Some spaces, like the Visagie farm or the orphanage, are less overtly controlling than others, operate within an economy of altruism, and are expressed through a politics of care. But Michael K remains unconvinced by these charitable offerings: 'I have escaped the camps; perhaps, if I lie low, I will escape the charity too.'[51] Michael K seems to recognize the nature of these places. To him, all social relations are a form of oppression.

The second method individuals in the novel deploy to make sense of Michael K is by directly soliciting narratives from him. They ask about his life, where he has been, what his plans are. These figures are hungry for the words that will rationalize and explain Michael K's somatic and behavioural difference. Language is analogous to food in *Life & Times of Michael K*, with its orally disfigured protagonist experiencing a problematic relationship not only with eating but also with speaking. Michael K remains silent in the face of questioning, unable or unwilling to produce the sought-after confirmation of his identity that his interlocutors hunger for. The belief that words can take the place of food goes back as far as the Old Testament. The Book of Revelation depicts an angel of the Lord urging the narrator to eat the sacred book, emphasizing the transformative qualities of eating: 'it shall make thy belly bitter, but it shall be in thy mouth sweet as honey.'[52] Exploring the links between food and language, Deleuze and Guattari state:

> The original territoriality of the mouth, the tongue and the teeth is food. By being devoted to the articulation of sounds, the mouth, the tongue and the teeth are deterritorialized. So there is a disjunction between eating and speaking [...] to speak [...] is to starve.[53]

Here Deleuze and Guattari posit a mutually exclusive relationship between food and words, claiming that one necessarily takes the place of the other. The body is compromised by speech, its boundaries invaded and its integrity weakened. Following this logic, if Michael K were to speak, it would cause a further assault on a body already subject to a number of controlling and classificatory practices, many of which – like the work camp – he cannot escape. Michael attempts to preserve the boundaries of his bodily self by refusing to speak. In this instance, silence might be interpreted as an act of resistance, or it may be read as a lack of agency. Michael's hunger contains ambivalences that he fails to clarify or define. Michael K's body is a contested space; characters deploy searching queries in an attempt to rhetorically situate him within a narrative that is sense-making and obeys the rationality of the state and the contemporaneous civil unrest.

In response to many characters' interrogations of him, Michael K wonders: 'This was evidently a code for something, he did not know what [...] was he expected to say something?'[54] Michael K's anxiety about being swallowed up by others' hunger is echoed in his trepidation when speaking to them: 'The words, whatever they stood for, accusation, threat, reprimand, seemed to K to smother him.'[55] A familiar fear of engulfment is enacted here; his reluctance to engage in speech obeys the same rationale as does his selective aversion to food. Upon his initial arrival at Jakkalsdrif, Michael is introduced to other inmates in the camp, and he possesses a dim understanding that he is expected to produce some explanation or history of himself: 'There was a silence. Now I must speak [...] thought K, so as to be complete, so as to have told the whole story. But he found that he could not, or could not yet.'[56] The inclusion of the word 'complete' resonates with an idea of satiety as well as narrative closure – but Michael fails to satisfy either of these hungers. If Michael responds to external questioning, he allows himself and his body to be interpellated as a subject, particularly given the already politically charged spaces that he finds himself in – a prison camp, a work camp, a hospital, etc. The words of others seem smothering and invasive, so silence is conceived as resistance. By shoring up the borders of his body, he expresses a subversive politics of freedom:

> Always, when he tried to explain himself to himself, there remained a gap, a hole, a darkness before which his understanding baulked, into which it was useless to pour words. The words were eaten up, the gap remained. His was always a story with a hole in it: a wrong story, always wrong.[57]

The use of the word 'baulk' here echoes Michael K's previous fears of becoming a 'beast at the shambles' earlier in the novel. Michael K is unable to fill the gaps in his narrative to any audience's satisfaction – ours or that of the other characters in the novel. The hunger perpetually remains. Although his words are eaten up, the hunger for completion – for satisfaction – proves impossible to sate or overcome.

'All of Coetzee's novels,' Stephen Watson argues,

> contain passages that express a great longing for history. They are unfailing in their desire for a world of event, for a narrative in which there is direction and purpose, a story which has a beginning and an end, in which character has some continuity in time.[58]

This hunger is expressed by both the reader and the characters within the novel, who often serve as a sort of avatar for our own narrative appetites. The hunger is for form – a recognizable and comforting teleological shape that produces a satisfactory, neat ending. The imagined temporal structure of narrative, it is conventionally understood, is a utopian one. In *Marxism and Form*, Jameson quotes Ernst Bloch to explain that textual narratives obey a unitary rationale, one that moves from chaos to harmony:

> Every great work of art, above and beyond its manifest content, is carried out according to a latency of the page to come, or in other words, in the light of the content of a future which has not yet come into being, and indeed of some ultimate resolution as yet unknown.[59]

This drive for narrative completion is comparable to the phenomenology of consumption; satisfaction, or fullness, is sought in the denouement of the text. But as the assumed utopian form of narrative presents a false teleology, mirroring the imagined grand historical narratives of Modernity and European history, so does the novel play with its readership's normative textual appetites. The linear and totalizing quality of history is interrogated through Michael K's evasions, by the plot of the novel, and by the form of the novel – allegorical or otherwise. Like hunger itself, the text never truly satisfies with an objective, or 'truthful,' conclusion.

Michael K's behaviour can be read as an attempt to perform the idealized 'closed' body – a typically modern western somatic ontology. Closed body practices embody power and control by separating the body from those around them. Susan Bordo states that the transition from open to closed bodies occurred during the European Renaissance – when identity went from being located in the world or community to being located in the interior – the mind.[60] The ideal of the closed body is what Michael K attempts to replicate – self-sufficiency as freedom. The closed body and its associated practices are thus closely allied to the Cartesian model of the self. As cultures became more individuated, somatic practices focused on bodily self-control – biopower. Michael K's efforts exemplify these ideas, and we can trace them to European ontologies, and find their evidence in the colonial logic of the South African state. As such, the closed body ideal is replication of the master's methods and language, and the ambivalences – and failures – contained within Michael K's hunger can be understood through his problematic adoption of this discourse. He cannot escape the overdeterminations of his body any more than he can escape society itself. His silence may be conceived as a partial success, at least an attempt at moving beyond the burden of signification his body is situated in, but ultimately the silence seems to foreclose the possibility of saying something else, speaking through the body in a new, radical register. Nevertheless, Michael structures his acute hunger events using the same Cartesian discursive logic, and in the following section I explore the potential successes and failures of these moments.

Body and Mind

Michael K experiences two separate occasions of self-isolation in the novel, and both are a response to the repressive appetites portrayed in the book. Both are instances of hunger cut short by the invasion of unwanted outsiders who force him into a regulated environment: first when he is captured and recruited into the Jakkalsdrif work camp, and a second time when he is forced into medical care at the Kenilworth rehabilitation centre. Coetzee grants ample narrative attention to both episodes of isolation. In them, Michael K experiences a great reduction of hunger, and an unintelligibility present in both sections solicits an interpretative and deciphering reading of both instances. We are lured as readers (and even more so as critics) to understand Michael K's fierce determination to live in such an unconventional manner, in squalid and inhospitable circumstances that clearly fall outside the bounds of normative behaviour as understood by the novel. The food abnegation that he intentionally subjects himself to remains unexplained either by the narrative-free indirect discourse in which the majority of the text is written, or by the first-person voice of the medical officer. I read these periods of food refusal as comparable to a hunger strike, since they appear to be a form of protest, but I do not use the term hunger strike in its most conventional, political sense.

Overtly political hunger strikes rely on decipherability and clear aims to make sense. For political aims to be met, these aims must be clearly delineated and understood by both the striker and the audience that is witness to the strike. 'It is this silence that hunger strikers have to break if they intend to make their self-starvation readable as protest.'[61] However, Michael K is pointedly silent during these periods of food refusal; even when urged, he does not reveal the reason behind his ascetic practices. The frustrated medical officer in Chapter Two urges Michael K to reveal his motivations:

> There are hundreds of people dying of starvation every day and you won't eat! Why? Are you fasting? Is this a protest fast? Is that what it is? What are you protesting against? Do you want your freedom? If we turned you loose, if we put you out on the street in your condition, you would be dead within twenty-four hours. You can't take care of yourself, you don't know how. Felicity and I are the only people in the world who care enough to help you. Not because you are special but because it is our job. Why can't you co-operate?[62]

Michael K remains steadfastly silent. While this may appear to complicate my claim that Michael K is fasting in protest, I argue that his fast is indeed a form of dissent – although it both resembles and differs from the conventional political hunger strike. Political hunger strikes are rooted in the social context from which they emerge; the logic of the hunger strike is a subversion of pre-existing socio-political circumstances, but nonetheless it relies heavily on those circumstances (imprisonment, oppression, etc.) in order to enact and 'give sense' to the subversion. Thus, hunger strikers rely on language and representation to give shape and meaning to their transforming body. In contrast, Michael K appears to shun language and speech entirely. His is a protest against the body itself – the body as it is constructed by colonial discourse and mediated by the policing nation state. Paradoxically, the protest's first casualty is the body itself, which is so integral to the coherency of the self – and the very grounds of the protest. Michael K's protest is predicated upon a central contradiction: he is attempting to claim ownership of his own body by asserting his own meaning upon it, by secreting it away from the interpolating mechanisms of the state, and by occupying a state of incomprehensibility, which can all be read as a political response to oppressive colonial power. The attempt to somehow move beyond language – or, at least, the language of the colonizers – is an assertion of agency. Michael K protests against the very notion of intelligibility itself. His rejection of both words *and* food indicates a protest against forms of narrativization, against the discursive tendencies that attempt to render his starving body knowable and containable. His protest is a logical rejoinder to these discursive pressures, but whether his protest is successful is subject to ambivalences. His silence can be read as productive – an assertion of the somatic, the deployment of a dissenting form of speech in the form of the starving body. In his refusal to bend to the rational ideologies of the colonial state, Michael articulates a politics of illegibility, whose meanings can be read as transcending the logic of the colonial state. The structure of Michael K's strike is a Foucauldian dividing practice – a self-policing form of biopower.

Michael K is subject to the three different forms of 'objectification of the subject' that Foucault outlines in his work. First is 'dividing practices' – 'the subject is objectified by a process of division either within himself or from others'.[63] This form of objectification is a closed body practice and was considered in the previous section, whereby Michael's body is subject to spatial control and imprisonment, be that in a work gang or a medical centre. Second, Michael K is also objectified as a subject by Foucault's 'scientific classification' – 'the objectivizing of the productive subject, the subject who labours, in the analysing of wealth and of economics. Or [...] the objectivizing of the sheer fact of being alive in natural history or biology.'[64] Michael's stay at the Kenilworth, under the medical officer's care, falls into this category. There is a sense that cataloguing and control of the subject may be achieved by empirical classificatory processes, and the medical officer attempts to do this repeatedly. According to Paul Rabinow, Foucault's final form of objectification is 'subjectification':

> [T]hose processes of self-formation in which the person is active [...] this self-formation has a long and complicated genealogy; it takes place through a variety of operations on [people's] own bodies, on their own souls, on their own thoughts, on their own conduct. These operations characteristically entail a process of self-understanding buy one which is mediated by an external authority figure, be he confessor or psychoanalyst.[65]

It is this final mode of objectification that I wish to consider in more depth, as Michael K's hunger strike can be read productively through this model of subjectification. His hunger strike is an attempt at resistance, but one of the prevailing problematics of his strike is that it is a form of biopower whose forms are borrowed from the repressive state ideology that controls the landscape of the novel. The hunger strike is a response to the contradictions of the subaltern body as it is positioned within colonial discourse. The subaltern body internalizes the strictures of the colonial state to structure the resistance of hunger and is entrapped within them. Any action that may be construed as further styling, policing, mediating, and controlling the body – even one carried out by Michael K himself – might be read as an act of colonial violence.

Both periods of Michael K's food abnegation in the text share common characteristics, and in each Michael K retreats into interiority, and consequently minimalizes the material needs of his body. The narrative exposition preceding both events invites equally congruous readings. As the narrative has followed Michael K through various instances of exploitation and invasion by others, and this has been interpreted through the terms of hunger – whether for words or for food – it follows logically that Michael would attempt to shut down or negate the pull of hunger by retreating from both food and words. In complete seclusion, he is not forced to consume the food of others, nor to feed on their words in order to be 'complete.' He can be wholly closed off, 'like a stone,'[66] his inner world at peace from the prying questions and demands of the other.

Michael K's hunger is comparable to Nimi's from *The Inheritance of Loss*, as the two share some common features. Both characters attempt a rejoinder to the discursive demands that food and language place on their bodies by rejecting both foods and words. But Michael K's hunger strike goes one step further, as we have more access into his character than we do to the marginalized figure of Nimi.[67] Michael K empties the physical category of his subjectivity – his body – and emphasizes the ontology of the immaterial 'spirit' of the interior – the soul/mind. This involves a stringent dichotomizing between body and mind, which Michael first establishes when explaining his mother's cremated remains to a child: '"Did they burn her up?" asked the boy. K saw the burning halo. "She didn't feel anything," he said, "she was already spirit by then."'[68] Michael's efforts to starve meaning out of his body result in an effect that is common to the rhetorical descriptions of hunger strikes as they appear in this study. A movement towards disembodiment manifests in a lofty, airy, sublime experience comparable to flight and freedom:

> After the hardships of the mountains and the camp there was nothing but bone and muscle on his body. His clothes, tattered already, hung on him without shape. Yet as he moved about his field he felt a deep joy in his physical being. His step was so light that he barely touched the earth. It seemed possible to fly; it seemed possible to be both body and spirit.[69]

The image of flight is counter-constructed by the connotations of the body as material and heavy. By starving his body, Michael K attempts to physically disappear, erasing himself from the narrative and making obsolete the burdensome weight of his body, the site of his connection with others and the space of social co-option and oppression: 'He thought of himself not as something heavy that left tracks behind it, but if anything as a speck upon the surface of the earth.'[70] However, his efforts to disappear soon prove impossible, and this is explained by the narrator: 'When people died they left bodies behind. Even people who died of starvation left bodies behind.'[71] The illness that strikes Michael K during these periods of extreme food deprivation also goes some way towards connecting the mind to the materiality of the body. His physical, and more specifically alimentary, needs cannot be fully serviced by the meagre diet he allots himself.

Moreover, the necessity and desire to live are not wholly abandoned in isolation, and this is admitted as such: 'If he ate, eating what he could find, it was because he had not yet shaken off the belief that bodies that do not eat die.'[72] This is how Michael K remains anchored to the narrative. Even his starvation leaves evidence of the physical, however meticulously he attempts to leave no trace of his presence – exemplified during the second instance of isolation in the veld through his avoidance of building even a temporary shelter for himself. Ultimately, he remains to the last unable to escape the narrative, society, and the needs of the body itself.

The logic of the hunger strike falls along the familiar Cartesian lines of the body/mind divide. Here the 'spirit,' or the incorporeal component of subjectivity, is contrasted against the material self, or the body. The hunger strike doubles and re-performs the dichotomy of the savage, mute body (as represented by the native body) and the logical, idealized mind (as exalted in colonial discourse), and it attempts to discipline itself. It can be read as a dividing practice, one committed by the self upon the self. This practice re-enacts the controlling biopolitics of the state – a subjectification of the self. Michael deploys the racist strategies of colonial ideology against himself. He polices and codifies his own body using his own mind, and this manoeuvre is sourced from the colonial discourse that structures the novel's narrative, as well as the literary tradition it is situated within. It is a logical strategy: it is a contradictory response to a contradictory condition – the colonized subaltern body sees itself as an object to be disciplined and contained, because it rearticulates the ideologies it has been indoctrinated with – the legacy of imperialism and racism. The colonized, native body (hated by itself, unable to find expression for itself other than in the overdetermined site of abject meaning heaped on to it by colonial ideology) – expresses dissent in the only language that is known to it. It does not, however, provide Michael K with the escape and relief he seeks – in attempting to recreate the ideal individuated 'closed' body of western philosophy, he is confronted by its failings, and its contingent nature – his body remains material, permeable, and mortal.

However, Michael K's hunger can be read productively as an ambivalent practice of dissent. His hunger is a violent act against the body and results in the disintegration of the self, damaging ultimately both body and mind. The representational and material are co-constitutive and interdependent. A deconstructive reading of the binary logic that underpins the hunger strike reveals the failure of hunger as a self-disciplinary practice – but it also asserts the agency of the material body. If body is read as not just the material self but also as a representation of the extratextual material world – a world, or history, outside the narrative – the book can be read as an assertion of the real, quite the opposite of Coetzee's critics who condemn his work as apolitical (nonetheless, one that is not arrived at easily through readings). We may read Michael K's materiality metaphorically, for that which escapes the bounds of discourse – an irrepressible materiality. Michael K's hunger deconstructs the body/mind binary, and through this exposes the epistemologies of prevailing colonial discourse, and how power is produced within them. The abject, savage body is a necessary component of colonial racial hierarchies – the uncivilized 'body' to be rationalized by the civilizing 'mind' of colonial power. This is thrown into sharp relief through reading Michael K's hunger, and as such invites us to interrogate power, knowledge, and the colonial subject. These are productive outcomes of a reading of hunger in this text.

By inserting the hungry body within South Africa's history of food insecurity, the body can tell various stories. Affirming racialized conceptions of the black body, Michael K's body can be read as the incompetent 'Cape Coloured' – unable to feed himself even when food is offered. This reading reinforces the cultural racism of the Apartheid state and renders the black body as backward and a reject of Modernity. But the starving body can also reveal in its emaciation the untold story of deprivation and lack in South Africa. That food is offered to Michael K, and that famine is largely a problem of the past, is no indication that the nation's oppressed are well fed. Instead, they are 'starving on a full stomach'[73] – an embodied history of ongoing, endemic poverty that keeps bodies alive but obliterates the soul. Michael K's hungry body can be read as a failure of the colonial state's reliance on the universalizing ideals of Modernity. By situating Michael K's body within this material history of hunger and lack, the hunger contained in the novel can be used to challenge the orthodoxy of colonial state racism, and its reverence of the free market. Michael K is not hungry because he does not know how to eat. He is hungry because he does not know how to eat within the confines of racial and economic inequality. By reading Michael K's hunger carefully, we can produce multiple readings, some of which disrupt colonialism's discursive power.

The politics of refusal contained in the protest can also be read as a rejoinder to the primitive 'animal-like' nature of the colonized body. By rejecting food when it is offered, the politics of sympathy is reversed. Instead, the body – divested of agency within the context of colonial control – enacts a self-destruction whereby at the very least agency can be asserted on life itself. This can be read as a response to bio-control, as the body subverts the image of the suffering oppressed, and intentionally figures this image upon the body itself – authored to be read in a way that counters the charitable/normalizing administration of the racist state. Or it can be taken one step further and interpreted as a response to necropolitical control. Mbembe states that 'the ultimate expression of sovereignty resides, to a large degree, in the power and capacity to dictate who may live and who must die.'[74] Within the racist politics and policies of the Apartheid state, and in the context of the widespread food insecurity in South Africa's history, this necropolitical reading is a useful extension of Foucault's work. Situated within the civil unrest of the novel, the necropolitical conditions of Apartheid are magnified and legitimized with further violence. The right to die, in this context, is wrestled from necropolitical control in the logic of the hunger strike. Michael K's body exposes the hypocrisies of the paternalistic colonial government, and does so through the startling image of the starving body – a body that might be read as evidence of a lack of cultural competence, but that can also be read as an indictment of the failures of the state. This body is hypervisible in its horrors, so the message is writ large on the body for all to see, even as it disappears into starvation. This demonstrates 'a politics of refusal that subverts the logics of recognition, empathy and suffering liberal rights discourses rely on.'[75] The abject starving body – neither alive nor dead, a walking corpse – pushes on the limits between life and death, both making material the histories and suffering of the wretched of the earth, yet refusing to participate in the economy of sympathy and charity that would continue to re-entrench him within the civilizing discourse of the Apartheid state. Michael K may be read as savage and backward, but his asceticism demonstrates mastery over his body, by some cognitive process that we are never fully privy to, but that nonetheless distances him from the purported uncontrollable, animal appetites of the colonial subject.

Hungry Readers

Despite shying away from other characters within the novel, Michael K remains rooted in the narrative itself by our own interpretive gaze. He comes into being, in fact, through our readership. A relationship of desire and hunger exists between reader and text. The reader is hungry for narrative intelligibility and consumes the narrative that constitutes Michael K's body, to render him intelligible. This is transposed from the dynamics of consumption that is contained within the plot of the novel. Now we are faced with our own voraciousness, the promise of narrative closure. The process of reading mimics the act of eating in the sense that it is a continuously destabilizing yet simultaneously palliative drive, one that nears fulfilment and stability, but reveals in its own process the impossibility – and violence – of such desires. As the novel itself states: 'The gaps always remained.'[76] The 'hunger' enacted in this reading of the book now implicates the reader-as-subject, who is drawn into a dependent relationship with Michael K's body-as-subject. The greater the unintelligibility of Michael K's narrative, the greater the effect of destabilization on the reader. This can be a productive destabilization, but only within the context of acknowledgement of our own reading practices as a desire for mastery and perfect understanding. We are here faced with our own Will to Knowledge – and our own 'open,' fluid, destabilized body. And as our hungry reading references the consumptive desire towards Michael's body that we have already seen displayed by the book's characters (particularly the medical officer), we are simultaneously faced with the possible colonizing nature of our own interpretive gaze, as well as our literary traditions and practices. This is particularly politically useful when considering the English-speaking, academic, and western audience that makes up the novel's readership.

The reader spends the course of the novel searching for clues to render Michael K's narrative and character more palatable; he/she becomes involved in a relationship, through the act of reading, that strikes at the very heart of readerly coherency. We wish to reach the point of Michael K's narrative completion and to discover that this threshold contains the answers we have been directed to crave (both by conventional standards of reading and by cues embedded within the text) in order to feel the stability of narrative completion – a completion that a realist narrative promises and an allegorical one tempts. But allegory moves the obligation of narrative logic from the text to the reader. Maureen Quilligan insists that allegory emphasizes the process of signification itself and therefore raises the reader's awareness 'of the way he creates the meaning of the text.'[77] *Life & Times of Michael K* is riddled with instances of confusion and a sense of oddness. This is due in large part to the lack of accessibility to its protagonist's interior. Michael K continuously engages in anomalous behaviour within the context of his own particular circumstances (which are hardly normative, given the emergency state of the country), and in *any* context: starving; living in isolation; the inability or unwillingness to share information, hopes, dreams, requests for help; and even the ability to thank strangers for kindnesses. All of his 'prodigious' activities contribute to the characteristic strangeness of the novel, and they remain inexplicable, the narrator providing none of the sought-after information. Even the free indirect discourse provided by the narrator affords little relief from this feeling of alterity, as the voice provided is often cryptic and only further confuses.

It is the reader's hungry gaze, his/her appetite for teleological satiation, that pins and locates Michael K in the narrative despite his best attempts to break free. The novel's narrator colludes in this process. A distance is created by the third-person narration. For instance: 'An hour later K was still sitting there, asleep, his mouth agape. Children, whispering and giggling, had gathered around him. One of them delicately lifted the beret from his head, put it on, and twisted his mouth in parody.'[78] The novel's narrator appears as though it is focalized through Michael K, but, as this scene demonstrates, when Michael K is unconscious, narration continues. The narrative style draws attention to the disconnect between Michael K's interior and itself. The text gestures back towards itself, revealing the artifice of its rhetorical techniques and for a moment suspending our immersion in the narrative – drawing attention to the reader's experience. These moments of dislocation draw attention to our hunger to gain access into Michael and interrupt an unproblematic trajectory of reading. There are no truly 'closed' bodies in the novel – or even outside of it. And as the Power/Knowledge of the novel is made plain to the reader, we are reminded of our collusion within it. Reading Michael K through this relational schema allows us to reorient our own position.

The reader's response to this disconnect is an interpretative one. Readers are tempted to know/consume Michael K to ascertain the true 'message' of the novel. As Attridge notes in his critical collection *J.M. Coetzee and the Ethics of Reading*, confusion resulting from oddities in the narrative tempts the reader to 'deploy reading techniques that will lessen or annul the experience of singularity and alterity – and this will usually involve turning the event into an object of some kind (such as a structure of signification).'[79] Our desire to consume Michael K is a disciplinary drive. This desire mimics the hunger of characters in the novel, especially the medical officer. 'Michaels means something,'[80] the medical officer continues to insist:

Michaels means something, and the meaning he has is not private to me. If it were, if the origin of this meaning were no more than a lack in myself, a lack, say, of something to believe in, since we all know how difficult it is to satisfy a hunger for belief with the vision of times to come that the war, to say nothing of the camps, presents us with, if it were a mere craving for meaning that sent me to Michaels and his story, if Michaels were no more than what he seems to be (what you seem to be), a skin-and-bones man with a crumpled lip (pardon me, I name only the obvious), then I would have every justification for retiring to the toilets behind the jockey's changing-rooms and locking myself into the last cubicle and putting a bullet through my head.[81]

In this passage, the medical officer affirms the insatiable hunger for meaning. Michael K's refusal to conform to the logic of the state terrifies him. His interpretative desire doubles the reader's and brings into relief the discomfort of the text's engagement with intelligibility. A lack of meaning as represented by Michael K's body destabilizes the teleology of the novel, and so of historical metanarratives, which highlights both the medical officer's anxieties and our own. The lack, or 'hunger,' described by the medical officer refers to a desire to believe that representation – in this case, one of Michael K as a meaning-bearing sign – is not arbitrarily constructed upon a historical reality that may, in the end, be meaningless and devoid of universal truths. Thus, Coetzee's claim that allegory can 'usurp history's discursive position' is realized.

Attridge notes that these moments of unintelligibility will vary from reader to reader, as individual readership is determined by a wholly unique set of factors ranging from the social to the personal. He advocates what he brands a 'literal reading' of the novel:

That is to say, in literary reading (which I perform at the same time I perform many other kinds of reading) I do not treat the text as an object whose significance has to be divined; I treat it as something that comes into being only in the process of understanding and responding that I, as an individual reading in a specific time and place, conditioned by a specific history, go through.[82]

Proposing such a reading advocates treating the text as an event, as opposed to an allegorical object that reveals in its various codifications certain extratextual truths (although Attridge acknowledges the automatic and simultaneous occurrences of such readings). Although one brings to the text their familiarity with a range of literary conventions that facilitate a lucid reading, Attridge notes that in Coetzee's writing there often occurs a sort of 'strangeness, a newness, a singularity, an inventiveness, an alterity in what I read.'[83] Attridge argues that from this point onward, there exist two options: to shut down the experienced strangeness by converting the experienced alterity into a known object (by deploying reading strategies that turn the oddness into some sort of allegorical signification), or to allow the strangeness of the reading experience to stand for itself – strange and unknowable. He argues for a reading that allows for the continual 'living' the text as a disruptive event, one that changes in every new reading. This kind of reading allows for an individualized reading experience and takes into account the difficulty of conveying such a unique and reader-dependent experience of 'otherness' in a critical reading. This interpretive approach permits something to escape the intelligibility of the narrative – it admits that something can escape the limits of discourse.

This idea is repeated by the medical officer – the author-as-avatar voice in the novel:

> Your stay in the [Kenilworth] camp was merely an allegory, if you know the word. It was an allegory – speaking at the highest level – of how outrageously a meaning can take up residence in a system without becoming a term in it. Did you notice how, whenever I tried to pin you down, you slipped away?[84]

The medical officer affirms the indecipherability of Michael K's motivations. As Attridge argues, this ambiguity is intentionally left to counter the act of directed, interpretative reading; in other words, it is 'Against Allegory.' If the otherness that Attridge describes is left alone and not incorporated into a singular allegorical meaning, a productive reading is possible, and one that limits discursive power:

> The point I wish to make is that allegorical reading of the traditional kind has no place for this uncertainty and open-endedness, this sense that the failure to interpret can be as important, and quite as emotionally powerful as success would be.[85]

This method of reading asks the reader to disrupt the consuming gaze – our Will to Knowledge. This allows the space for otherness – an otherness that breaks through our systems of knowledge.

A hierarchical binary is created between an unknowable, frightening, and mute material category and the discursive significations that bring order, stability, and meaning to it. This binary is restated in the Cartesian binary created between body and mind that appears in the novel, in which Michael's Othered body is the material site that is policed, corrected, civilized, and rendered incomprehensible by the rationality of the Apartheid state. Attridge claims that each moment of meaninglessness 'opens a space for the other.'[86] The drive to understand Michael is indicative of traditional literary readings that attempt to decipher this unknowability allegorically: 'In order to move to parallels outside the world of the book, many of the rich and sometimes apparently quite contingent details of the text have to be ignored.'[87] Atwell echoes these sentiments:

> To the last K remains his own person: in refusing to be imprisoned in any way, either in the literal camps or in the nets of meaning cast by those

who follow after him, he becomes [...] a principle of limited, provisional freedom, a freedom located in the act of writing.[88]

Both Atwell's and Attridge's readings legitimate the possibility of indecipherability-as-freedom in the *Life & Times of Michael K* – something that escapes the confines of discourse. This provides the basis of reading the novel as an act of political intervention and is a productive imagining of the force of hunger. We relate to Michael not through the strictures of oppressor/oppressed, or via the power differentials of sympathy, but are placed next to him. Instead, we align with the perspective of oppressor, and this forces a process of self-interrogation.

The possibility of a freedom that might supersede the strictures of discourse faces opposition from poststructural readings of the novel. Far from opening a 'space for the other,' such readings instead locate Michael K's unintelligibility as less of an avenue toward radical extra-linguistic representation and more as a term that is lodged inside the discursive logic of the text. Michael K's inscrutable body is situated in a larger sense-making discursive matrix of textuality – a term of illegibility and illogic that is a necessary component that upholds the text's internal structures of power. Discourse's totalizing power renders Michael K's irrationality-as-freedom an impossibility; instead, if anything, the text can be read as appropriating his silence, his protest, his alterity and hunger as a fixed discursive boundary between the logic of Modernity and the savagery of the colonial body – contained within the strictures of textuality. This critical framing limits the text's accommodation of political freedom as mentioned by Atwell and Attridge, instead confirming that 'discourse determines the limits of thought within a society, that no subject can exist outside of discourse, and that no form of resistance is therefore possible from this imagined extra-discursive position.'[89] Michael K's body is encircled by the novel's discourse. Its materiality fails to break through the representational and textual traps that threaten to engulf it within the novel, and, ultimately, its limited agency as a site of a radical articulation that transcends discursive logic – proving the novel's political possibilities – is curtailed by the text itself.

It is difficult to deny the totalizing power of discourse, or the critical authority of these kinds of poststructural readings – but there may be a way of understanding Michael K's protest and body that leaves the absolute limits of discourse intact, but still interrogates the place of materiality and the body within the larger power structures of the text. Affect theory and associated theories of the body provide such possibilities. Elspeth Probyn states:

> The body is always outlined by the structures of a given social formation, and it is always lived at a tangible level. It must be theorized as such, caught within the tensions of both 'structure' and 'feeling.' Thus the body as simply a re-creation of the image of its feeling is displaced–the spoken image of the body can never be at one (in simple equivalence) with an untheorized notion of experience. At the same time, we can acknowledge the limitations of the body as a disembodied and ideologically interpellated subject position.[90]

Probyn advocates for reading multiplicities in the body and thus revitalizes it beyond its representational constrictions. More than an inert and flattened discursive category, Probyn highlights the body's relational qualities, and this in turn empowers the body with not just more agency to interrogate its fixed place in the binary logic of discourse, but also gestures to something that is beyond language's capacity to wholly represent:

> integral to a body's perceptual becoming (always becoming otherwise, however subtly, than what it already is), pulled beyond its seeming surface-boundedness by way of its relation to, indeed its composition through, the forces of encounter. With affect, a body is as much outside itself as in itself – webbed in its relations – until ultimately such firm distinctions cease to matter.[91]

Strictly poststructuralist readings of *Life and Times of Michael K* overemphasize the binary logic of the body/mind and material/representational, and fail to imbue the body with the fluid agency that Probyn emphasizes, even within the representational totality of discourse.

These theories helpfully dovetail with 'open' theories of the body that again call attention to the ways in which lived bodies interact and continuously affect one another. Rather than conceptualizing bodies as closed physiological and subjective systems, bodies are open, 'participating in the flow or passage of affect, characterized more by reciprocity and co-participation than boundary and constraint.'[92] Lisa Blackman conceives of the body as a fluid process between self and other, and the haptic nature of these interactions. In these interactions, something might escape discourse.

> Haptic, or affective, communication draws attention to what passes between bodies, which can be felt but perhaps not easily articulated. The more non-visual, haptic dimensions of the lived body distribute the idea of the lived body beyond the singular psychological subject to a more intersubjective and intercorporeal sense of embodiment This is embodiment as intercorporeality.[93]

This intercorporeality is what gives the hunger strike its power. The hunger strike is never simply about the singular, closed body, asserting a body technique upon itself in insolation. The failures of Michael K's hunger demonstrate the impossibility of such a stance – although, of course, we must be reminded that the haptic or affective as they are represented within a work of literature are necessarily mediated by text.

But there is more than one type of body to consider when thinking through the somatic – there is the body concept, as well as the one that might escape the confines of language as affect. As Probyn states: 'In stressing the epistemological weight of the doubleness of the body, I will argue that we can use the body in two registers: in one, the body is positioned as concept, and in another, we can think the body within strategies of enunciation.'[94] Whether there is something affective and textually uncontainable about the transformation Michael K's bodily protest has on the medical officer, or the reader of the novel, will remain ambiguous. Affect theory's legibility within a literary context must be situated within language itself and is necessarily mediated by discourse when both reading affect and writing about affect. Affect is hard to contain and quantify, which makes it both an effective platform from which to approach the limits of discourse and a difficult thing to speak and write about. However, the body-as-concept can also disrupt on a discursive level. It can unsettle the body of the other, when it interrogates the limits of its own formations, if we remind ourselves of the multiple, relational bodies that circulate within the discursive field. '[A]ffects are forms of encounter; they circulate – sometimes ambivalently but always productively – between and within bodies (of all kinds), telling us something important about the power of affect to unravel subjectivity and modify the political body.'[95] In the flattened terrain of the text, the body's interactions with its various discursive oppositions – other bodies, the mind, the world, the text – can transform it, and inject some level of agency into embodied practice.

The 'closed' body Michael K tries to achieve is impossible, as it is constructed by and within our reading of him. It is a reminder that bodies are never totally closed; rather, they exist in a system of interrelations that are malleable and contingent. They are permeable – like the semiotics of language, the sign of the body cannot exist in isolation. Michael K's always-open mouth signifies the openness of the body, and gestures toward an allegorical reading of the text that opens up multiple historical readings, empowering the reader to apply readings that may politicize the text – or not. Certainly, the book is not without its ambivalences, as is demonstrated through the failures and successes of Michael K's hunger. However, through an interrogation of the body and its possibilities – discursively, politically, and affectively – this chapter reads the novel as a potential site of writerly agency, rather than Roland Barthes' readerly text – 'In other words, the multiple truths of the body force us beyond a theory lodged in duality'[96] – and allows us to use the body to read against the fixed divisions of mind/body, reason/emotion, subject/object, representational/discursive and colonized/colonizer.

Notes

1 Elleke Boehmer, 'Transfiguring: Colonial Body into Postcolonial Narrative', *NOVEL: A Forum on Fiction*, 26.3 (1993), 268–277 (p. 269).
2 See Michael Vaughan, 'Literature and Politics: Currents in South African Writing in the Seventies', *Journal of Southern African Studies*, 9.1 (1982), 118–138; Paul Rich, 'Apartheid and the Decline of the Civilization Idea: An Essay on Nadine Gordimer's *July's People* and J.M. Coetzee's *Waiting for the Barbarians*', *Research in African Literatures*, 15.3 (1984), 365–393; Peter Knox-Shaw, '*Dusklands*: A Metaphysics of Violence', *Commonwealth Novel in English*, 2.1 (1983), 65–81; and Nadine Gordimer, 'The Idea of Gardening', *New York Review of Books*, 3.6 (1984).
3 Z.N., 'Much Ado about Nobody. Review of *Life & Times of Michael K*, by J.M. Coetzee', *African Communist*, 97 (1984), 101–103 (p. 102).

4. J.M. Coetzee, *The Novel Today*, Container 64, Folder 7, Early Works, J.M. Coetzee Papers (Harry Ransom Center, University of Texas at Austin, 1987).
5. Dominic Head, 'The (Im)possibility of Ecocriticism', in *Writing the Environment: Eco-criticism and Literature*, ed. by Richard Kerridge and Neil Sammells (New York: Zed, 1998), 27–39.
6. It does this by paying close attention to Michael K's relationship to his gardening duties, using this as a platform for reorienting the relationship between man and nature as a vital and powerful justificatory source for ethical eco-awareness.
7. Anthony Vital, 'Toward an African Ecocriticism: Postcolonialism, Ecology and *Life & Times of Michael K*', *Research in African Literatures*, 39 (2008), 88–94 (p. 92).
8. Derek Attridge, *J.M Coetzee and the Ethics of Reading: Literature and the Event* (Chicago, IL: Chicago University Press, 2004), p. 32.
9. David Atwell, *J.M. Coetzee: South Africa and the Politics of Writing* (Berkeley: University of California Press, 1993), p. 2.
10. Vaughan, p. 128.
11. Michael Chapman, 'The Writing of Politics and the Politics of Writing on Reading Dovey on Reading Lacan on Reading Coetzee on Reading …(?)', *Journal of Literary Studies*, 4.3 (1988), 327–341 (p. 338).
12. Alys Moody, *The Art of Hunger: Aesthetic Autonomy and the Afterlives of Modernism* (Oxford: Oxford University Press, 2018), p. 158.
13. Jarad Zimbler, 'For Neither Love nor Money: The Place of Political Art in Pierre Bourdieu's Literary Field', *Textual Practice*, 23.4 (2009), 599–620 (p. 612).
14. Louise Bethlehem, '"A Primary Need as Strong as Hunger": The Rhetoric Urgency in South African Literary Culture under Apartheid', *Poetics Today*, 22.2 (2001), 365–389 (p. 376).
15. Ibid., p. 378.
16. Ibid., p. 378.
17. Stephen Slemon, 'Modernism's Last Post', in *Past the Last Post: Theorizing Post-Colonialism and Post-Modernism*, ed. by Ian Adam and Helen Tiffin (Calgary: University of Calgary Press, 1990), 1–12 (p. 7).
18. Aijaz Ahmad, 'Jameson's Rhetoric of Otherness and the "National Allegory"', *Social Text*, 17 (1987), 3–25.
19. Fredric Jameson, 'Third-World Literature in the Era of Multinational Capitalism', *Social Text*, 15 (1986), 65–88.
20. J.M. Coetzee, *White Writing: On the Culture of Letters in South Africa* (New Haven, CT: Yale University Press, 1988), p. 11.
21. Elena Gomel, 'The Poetics of Censorship: Allegory as Form and Ideology in the Novels of Arkady and Boris Strugatsky', *Science Fiction Studies*, 22.1 (1995), p. 89.
22. Judith Anderson, *Reading the Allegorical Text: Chaucer, Spenser, Shakespeare, Milton* (New York: Fordham University Press, 2008), p. 29.
23. Ibid., p. 5.
24. Moody, p. 157.
25. Terry Eagleton, 'Edible Ecriture', *Times Higher Education*, 1997, www.timeshighereducation.co.uk/features/edible-ecriture/104281.article [accessed 11 November 2020].
26. Moody, p. 171.
27. J.F. Herbst, 'The Administration of Native Affairs in South Africa', *African Affairs*, XXIX.CXVII (1930), 478–489.
28. Francis Wilson and Mamohela Ramphele, *Uprooting Poverty: The South African Challenge: Report for the Second Carnegie Inquiry into Poverty and Development in Southern Africa* (New York: W.W. Norton & Co., 1989), p. 101.
29. Diana Wylie, *Starving on a Full Stomach: Hunger and the Triumph of Cultural Racism in Modern South Africa* (Charlottesville: University of Virginia Press, 2001).

30 Human Sciences Research Council, *Food Security in South Africa: Key Policy Issues for the Medium Term*. Position Paper (Pretoria: Human Sciences Research Council, 2004).
31 Wylie, p. 4.
32 J.M. Coetzee, *Life & Times of Michael K* (London: Vintage, 2004), p. 7.
33 Ibid., p. 142.
34 Ibid., p. 14.
35 Ibid., p. 10.
36 Ibid., p. 14.
37 Ibid., p. 40.
38 Ibid., p. 98.
39 Ibid., p. 135.
40 Boehmer, p. 269.
41 Coetzee, *Life & Times of Michael K*, p. 150.
42 Ibid., p. 5.
43 Ibid., p. 7.
44 Ibid., p. 7.
45 Ibid., pp. 8–9.
46 Ibid., p. 36.
47 Ibid., p. 35.
48 Ibid., p. 71.
49 Maud Ellmann, *The Hunger Artists: Starving, Writing and Imprisonment* (London: Virago Press, 1993), p. 29.
50 Michel Foucault, *Discipline and Punish: The Birth of the Prison* (New York: Vintage, 1995), p. 136.
51 Coetzee, *Life & Times of Michael K*, p. 182.
52 *The Holy Bible Containing the Old and New Testament, Authorized King James Version* (Victoria: Bible Protector, 2007), Revelation 10:9.
53 Gilles Deleuze and Félix Guattari, quoted in Mladen Dolar, 'Kafka's Voices', in *Lacan: The Silent Partners*, ed. by Slavoj Žižek (London: Verso, 2006), p. 332.
54 Coetzee, *Life & Times of Michael K*, p. 31.
55 Ibid., p. 64.
56 Ibid., p. 79.
57 Ibid., p. 110.
58 Stephen Watson, 'Colonialism and the Novels of J.M. Coetzee', in *Critical Perspectives on J.M. Coetzee*, ed. by Graham Huggan and Stephen Watson (London: Macmillan Press, 1996), 13–36 (p. 32).
59 Ernst Bloch, quoted in Fredric Jameson, *Marxism and Form* (Princeton, NJ: Princeton University Press, 1971), p. 149.
60 Susan Bordo, *The Flight to Objectivity: Essays on Cartesianism and Culture* (Albany, NY: SUNY Press, 1987).
61 Ellmann, p. 18.
62 Coetzee, *Life & Times of Michael K*, p. 145.
63 Michel Foucault, *The Foucault Reader*, ed. by Paul Rabinow (New York: Pantheon Books, 1984), p. 8.
64 Ibid., p. 9.
65 Ibid., p. 11.
66 Coetzee, *Life & Times of Michael K*, p. 135.
67 Still, it is important to note that this internal dialogue is delivered in the form of a third-person narrator, which therefore acknowledges and emphasizes again the indecipherability of Michael's true motivations and characterization.
68 Coetzee, *Life & Times of Michael K*, p. 48.
69 Ibid., p. 102.
70 Ibid., p. 97.
71 Ibid., p. 94.

72 Ibid., p. 101.
73 Wylie, p. 237.
74 Achille Mbembe, 'Necropoltics', *Public Culture*, 15.1 (2003), 11–40 (p. 11).
75 Michelle Pfeifer, '*Becoming Flesh*: Refugee Hunger and Embodiments of Refusal in German Necropolitical Spaces', Citizenship Studies, 22.5 (2018), 459–474 (p. 459).
76 Coetzee, *Life & Times of Michael K*, p. 110.
77 Maureen Quilligan, *The Language of Allegory: Defining the Genre* (Ithaca, NY: Cornell University Press, 1979), p. 28.
78 Ibid., p. 50.
79 Attridge, p. 40.
80 The medical officer mistakenly refers to Michael K as 'Michaels' over the course of his first-person narration of him. This misreading has the effect of reasserting the illegibility of Michael K's character and also highlights the subjective nature of reading itself.
81 Coetzee, *Life & Times of Michael K*, p. 165.
82 Attridge, p. 39.
83 Ibid., p. 40.
84 Coetzee, *Life & Times of Michael K*, p. 166.
85 Attridge, p. 48.
86 Ibid., p. 64.
87 Ibid., p. 48.
88 Atwell, p. 92.
89 Moody, p. 187.
90 Elspeth Probyn, 'This Body Which Is Not One: Speaking an Embodies Self', *Hypatia*, 6.3 (1991), 111–124 (p. 117).
91 Gregory J. Seisworth and Melissa Gregg, 'An Inventory of Shimmers', in *The Affect Theory Reader*, ed. by Gregory J. Seisworth and Melissa Gregg (Durham, NC: Duke University Press, 2010), 1–26 (p. 3).
92 Lisa Blackman, *Immaterial Bodies: Affect, Embodiment, Mediation* (Thousand Oaks: SAGE Publications, 2018), p. 12.
93 Ibid., p. 12.
94 Probyn, 'This Body Which Is Not One', p. 111.
95 Caroline Williams, 'Affective Processes Without a Subject: Rethinking the Relation Between Subjectivity and Affect with Spinoza', *Subjectivity*, 3.3 (2010), 245–262 (p. 246).
96 Espeth Probyn, 'Theorizing Through the Body' in *Women Making Meaning: New Feminist Directions in Communication*, ed. by Lana F. Rakow (London: Routledge, 2017), 83–99 (p. 98).

Bibliography

Anderson, Judith H., *Reading the Allegorical Text* (New York: Fordham University Press, 2008)

Ahmad, Aijaz, 'Jameson's Rhetoric of Otherness and the "National Allegory"', *Social Text*, 17 (1987), 3–25.

Attridge, Derek, *J.M. Coetzee and the Ethics of Reading: Literature and the Event* (Chicago, IL: Chicago University Press, 2004)

Atwell, David, *J.M. Coetzee: South Africa and the Politics of Writing* (Berkeley: University of California Press, 1993)

Bethlehem, Louise, '"A Primary Need as Strong as Hunger": The Rhetoric Urgency in South African Literary Culture under Apartheid', *Poetics Today*, 22. 2 (2001), 365–389

Blackman, Lisa, *Immaterial Bodies: Affect, Embodiment, Mediation* (Thousand Oaks, CA: SAGE Publications, 2018)

Boehmer, Elleke, 'Transfiguring: Colonial Body into Postcolonial Narrative', *NOVEL: A Forum on Fiction*, 26. 3 (1993), 268–277

Bordo, Susan, *The Flight to Objectivity: Essays on Cartesianism and Culture* (Albany, NY: SUNY Press, 1987)

Chapman, Michael, 'The Writing of Politics and the Politics of Writing on Reading Dovey on Reading Lacan on Reading Coetzee on Reading …(?)', *Journal of Literary Studies*, 4. 3 (1988), 327–341

Coetzee, J.M., *Life & Times of Michael K* (London: Vintage, 2004)

Coetzee, J.M., *The Novel Today*, Container 64, Folder 7, Early Works, J.M. Coetzee Papers (Harry Ransom Center, University of Texas at Austin, 1987)

Coetzee, J.M., *White Writing: On the Culture of Letters in South Africa* (New Haven, CT: Yale University Press, 1988)

Dolar, Mladen, 'Kafka's Voices', in *Lacan: The Silent Partners*, ed. by Slavoj Žižek (London: Verso, 2006)

Eagleton, Terry, 'Edible Ecriture', Times Higher Education, 24 October 1997, www.timeshighereducation.co.uk/features/edible-ecriture/104281.article [accessed 11 November 2020]

Ellmann, Maud, *The Hunger Artists: Starving, Writing and Imprisonment* (London: Virago Press, 1993)

Foucault, Michel, *Discipline and Punish: The Birth of the Prison* (New York: Vintage, 1995)

Foucault, Michel, *The Foucault Reader*, ed. by Paul Rabinow (New York: Pantheon Books, 1984)

Gomel, Elena, 'The Poetics of Censorship: Allegory as Form and Ideology in the Novels of Arkady and Boris Strugatsky', *Science Fiction Studies*, 22. 1 (1995), 87–105

Gordimer, Nadine, 'The Idea of Gardening', *New York Review of Books*, 3:6 (1984)

Head, Dominic, 'The (Im)possibility of Ecocriticism', in *Writing the Environment: Eco-criticism and Literature*, ed. by Richard Kerridge and Neil Sammells (New York: Zed, 1998), 27–39

Herbst, J.F., 'The Administration of Native Affairs in South Africa', *African Affairs*, XXIX. CXVII (1930), 478–489

Human Sciences Research Council, *Food Security in South Africa: Key Policy Issues for the Medium Term*. Position Paper (Pretoria: Human Sciences Research Council, 2004)

Jameson, Fredric, *Marxism and Form* (Princeton, NJ: Princeton, 1971)

Jameson, Fredric, 'Third-World Literature in the Era of Multinational Capitalism', *Social Text*, 15 (1986), 65–88

Knox-Shaw, Peter, '*Dusklands*: A Metaphysics of Violence', *Commonwealth Novel in English*, 2. 1 (1983), 65–81

Mbembe, Achille, 'Necropolitics', *Public Culture*, 15. 1 (2003), 11–40

Moody, Alys, *The Art of Hunger: Aesthetic Autonomy and the Afterlives of Modernism* (Oxford: Oxford University Press, 2018)

Pfeifer, Michelle, '*Becoming Flesh*: Refugee Hunger and Embodiments of Refusal in German Necropolitical Spaces', *Citizenship Studies*, 22. 5 (2018), 459–474

Probyn, Elspeth, 'This Body Which Is Not One: Speaking an Embodies Self', *Hypatia*, 6. 3 (1991), 111–124

Probyn, Elspeth, 'Theorizing Through the Body', in *Women Making Meaning: New Feminist Directions in Communication*, ed. by Lana F. Rakow (London: Routledge, 2017), 83–99

Quilligan, Maureen, *The Language of Allegory: Defining the Genre* (Ithaca, NY: Cornell University Press, 1979)

Rich, Paul, 'Apartheid and the Decline of the Civilization Idea: An Essay on Nadine Gordimer's *July's People* and J.M. Coetzee's *Waiting for the Barbarians*', *Research in African Literatures*, 15. 3 (1984), 365–393.

Seisworth, Gregory J. and Gregg, Melissa, 'An Inventory of Shimmers', in *The Affect Theory Reader*, ed. by Gregory J. Seisworth and Melissa Gregg (Durham, NC: Duke University Press, 2010), 1–26

Slemon, Stephen, 'Modernism's Last Post', in *Past the Last Post: Theorizing Post-Colonialism and Post-Modernism*, ed. by Ian Adam and Helen Tiffin (Calgary: University of Calgary Press, 1990), 1–12

The Holy Bible Containing the Old and New Testament, Authorized King James Version (Victoria: Bible Protector, 2007)

Vaughan, Michael, 'Literature and Politics: Currents in South African Writing in the Seventies', *Journal of Southern African Studies*, 9. 1 (1982), 118–138

Vital, Anthony, 'Toward an African Ecocriticism: Postcolonialism, Ecology and *Life & Times of Michael K*', *Research in African Literatures*, 39 (Spring 2008), 88–94

Watson, Stephen, 'Colonialism and the Novels of J.M. Coetzee', in *Critical Perspectives on J.M. Coetzee*, ed. by Graham Huggan and Stephen Watson (London: Macmillan Press, 1996), 13–36

Williams, Caroline, 'Affective Processes Without a Subject: Rethinking the Relation Between Subjectivity and Affect with Spinoza', *Subjectivity*, 3. 3 (2010), 245–262

Wilson, Francis and Ramphele, Mamohela, *Uprooting Poverty: The South African Challenge: Report for the Second Carnegie Inquiry into Poverty and Development in Southern Africa* (New York: W.W. Norton & Co., 1989)

Wylie, Diana, *Starving on a Full Stomach: Hunger and the Triumph of Cultural Racism in Modern South Africa* (Charlottesville: University of Virginia Press, 2001)

Z.N., 'Much Ado about Nobody. Review of *Life & Times of Michael K*, by J.M. Coetzee', *African Communist*, 97 (1984), 101–103

Zimbler, Jarad, 'For Neither Love nor Money: The Place of Political Art in Pierre Bourdieu's Literary Field', *Textual Practice*, 23. 4 (2009), 599–620

4 Anorexic Fictions and Starving Histories in Tsitsi Dangarembga's *Nervous Conditions*

This chapter investigates the meanings of hunger in Tsitsi Dangarembga's 1988 novel *Nervous Conditions*. It considers hungry or starving bodies both within the novel and in the larger context of Zimbabwe's history of food scarcity. Forged within a history of famine and imperial violence, and subject to traditional Shona patriarchal control, a central character's body – a teenage hybrid subject named Nyasha – is read as a concentrated somatic representation of colonialism and its legacies, patriarchy, and intersectional oppression. Read in this context, Nyasha's food refusal may be interpreted as a transgressive response to the various ideologies at work on her body. Her body can also be read as an articulation of a problematic disavowal of material histories of hunger, lack, and colonial exploitation. These material histories respond to the misrepresentation of African hunger causality within popular western discourses, and the role of imperialism in this history. This chapter also considers the contradictory position of the hybrid colonial subject and how self-imposed hunger is used as a technique to overcome its paradoxes. The problem Nyasha experiences in stably representing her body, narrative, and identity is related to and vicariously experienced by her cousin Tambu – the focal narrator. This novel can be read as a Künstlerroman, with Tambu emerging from childhood to maturity in her transformed understanding of the socio-political context of race, gender, colonialism, Modernity, and development in Rhodesia. Her maturation allows her to write the narrative contained within the book.

The novel is set in the late 1960s against the backdrop of Rhodesia's civil war and the establishment of Ian Smith's white-minority-led UDI government.[1] Its realist narrative chronicles the mental and physical transformation of Tambudzai (or Tambu) Siguake – a teenage Shona girl living in rural Rhodesia. The narrative's discursive and geopolitical space is organized into a bi-cultural landscape that consists of two competing ideological hubs: Tambu's family home, the site of Shona traditionalism known as 'the homestead,' and the base of Rhodesia's Christian missionaries and location of Britain's colonial civilizing forces known as 'the mission.' The plot follows Tambu's ambitions of progression – through education – from the impoverished homestead to the promise of social and economic liberation as represented by the mission. Her own narrative trajectory parallels her cousin's: Nyasha is of a similar age but from a very different social

DOI: -4

and economic background. She has spent her adolescence in England until she is suddenly uprooted to Rhodesia by her father – the imposing patriarch Babamukuru. Both he and his wife Maiguru were educated in England, funded by a prestigious government-sponsored program, and now work and live in relative luxury with their two children at the mission; his household is a mixture of Shona traditional values and adopted western ideals.

Their daughter's resulting identity is hybridized: a mixture of western culture and Shona values. She struggles between the traditional expectations of Shona gender politics and the western values that she interprets as more progressive and modern. Critical work on the novel has tended to situate her within the context of this culture clash, particularly regarding the anomalous relationship she has with food – or her 'anorexia.'[2] Within the novel, food is realized as a powerful boundary-making symbol designating meaning and identity. Within the context of Rhodesia's past of imperial domination and hunger, bodies within this novel – particularly female bodies – become texts displaying the traces of a history of colonial domination. Throughout the course of the narrative, Nyasha displays what appear to be anorexic behaviours, wilfully rejecting the food on offer at her father's dinner table, which is realized as an intersecting site of complicit colonial and patriarchal oppression.

Tambu's construction of the mission as the means of realizing her ambitions of upward social mobility and freedom is complicated by Nyasha's protest against it, in the form of hunger. Through her starving body, Nyasha exposes the mission's pernicious ideological aims, and gestures toward its past colonial impacts. She chooses the intimate ground of her own body in order to enact a resistance to this narrative of exploitation; this response is staged upon the female Shona body – subject as it is to the intersections of colonial and patriarchal power. As the daughter of an authoritarian father who insists Nyasha behave like an appropriately 'decent' Shona woman, while also striving to achieve a western liberal ideal of Modernity, Nyasha is caught in a contradiction. Thus, Nyasha's identity is positioned in the double bind of gender and colonial oppressions. 'Anorexia nervosa is fundamentally about an identity crisis.'[3] Nyasha's body and protest can be read productively in so far as it embodies a central contradiction of the female postcolonial subject, and its position within colonial discourse and the colonial space. Her hunger is an analogous, contradictory response to an identity that exists in contradiction. The following section establishes the historical framework for this chapter by locating my reading of *Nervous Conditions* within a context of hunger and deprivation. I then consider the signifying logic of food, eating, and hunger within the novel and analyze Nyasha's 'hunger strike.' My analysis of the novel reads the Rhodesia of the novel through the socio-economic logic of food production and consumption. I situate Nyasha's hunger strike within this historical context, as well as considering the symbolic power of her hunger – a form of resistance that contains ambivalent ends, productive and destructive in equal measure. Nyasha chooses the Cartesian binary as a framework for her protest, and the disciplinary measures she self-imposes can be read as a rejoinder to the intersecting forces of gender inequality and colonial oppression.

Rhodesia, Hunger, and *Nervous Conditions*

The image of the starving African body is one of the most widespread contemporary representations of Africa. Pervading the western popular consciousness through imagery of starving children and emaciated adult frames, heavily circulated by the media, the problem of hunger in Africa has established itself as the most urgent call for international sympathy and aid. 'In its popular representations famine has emerged as the lodestar for African collapse and of the deep, enduring "crisis" of the post-independence project.'[4] While it is true that many Africans face food insecurity today – to varying extents – the ubiquitous representation of the starving African figure conceals the issue of causality: why has the black African body become synonymous with hunger?

> At present, most well-meaning people, encouraged by the mass media, see the hunger crisis as a result of the vagaries of nature [...] of such magnitude as to be virtually insoluble; as a scourge directly resulting from soaring birth rates in underdeveloped countries; or perhaps a question of laziness and lack of initiative on the part of the poor themselves.[5]

These implications transfer causal responsibility on to victims of hunger, applying theories about national and cultural underdevelopment or – more perniciously – essentialist and racist discourses about the primitive anti-Modern culture of African nations. In contrast to this, a report published by the Independent Commission on International Humanitarian Issues suggests an entitlement approach as a more productive means of understanding African food insecurity:

> The simple assumption that, if rains fail, as they have in recent years in much of Africa, less food will be grown and people will inevitably starve, may be a comfortable abdication of any human responsibility for what has happened. But it is a misleading simplification.[6]

Considering human involvement is imperative in framing the context of hunger and famine in Africa, as well as famine alleviation. Only through the metrics of food entitlement can we theorize an accurate understanding of why famines occur, and how they may be avoided. Scholarly interventions in the field of famine causation have overturned Malthusian explanations – such as his infamous theory of 'population checks'[7] – in favour of a food entitlement approach, a term most associated with the work of economist Amartya Sen. In Sen's theory, a lack of capital is the cause of hunger, as opposed to a general decline in the supply of food itself:

> A general decline in food supply may indeed cause him to be exposed to hunger through a rise in food prices with an unfavourable impact on his exchange entitlement. Even when his starvation is *caused* by food shortage in this way, his immediate reason for starvation will be the decline in his exchange entitlement.[8]

Or, it is not because of a failure in food production, but rather an individual's ability to afford – or be 'entitled to' – available resources that limits his/her access to food. This model emphasizes purchasing power as the means of understanding the root causes of hunger. This is particularly relevant in the historical context of Tsitsi Dangarembga's *Nervous Conditions*, because, in the case of hunger in Rhodesia, food availability was determined by high prices and unequal distribution rather than an overall failure in food supply.

Rhodesia has historically been troubled with food insecurity, experiencing no less than eight famines in the 60 years leading up to the historical period in which the novel is set, the late 1960s. In his book *Famine in Zimbabwe*, John Iliffe provides an account of the parameters, characteristics, and causation of these disruptions in food supply for the poorer inhabitants of the country. Pre-colonial Rhodesia's experience of hunger was characterized by relatively low mortality rates. Food shortages were generally caused by poor climate conditions, but the population adapted in response to food dearth by taking advantage of ecological diversity and stocking supplies. Food dearth rarely slipped into fatal famine conditions due to local trading relations and traditional defences against food scarcity. Due to these practices, not only were fatalities kept to a minimum, but shortages were localized to areas that were highly vulnerable to climate difficulties. In this environment, food dearth was a seasonal regularity in certain areas, and it was not surprising when a drought-driven crop failure caused a food shortage.[9]

With colonial interventions in the country came a change in the character of hunger. Land requisition, along with the introduction of a free-market economy, drastically reorganized the socio-economic landscape, and by 1922 it had become clear that the more traditional responses to famine had become mostly extinct. Market forces were in part responsible for the creation of food dearth, and migratory practices were favoured over customary foraging strategies. Land requisition by colonial forces across all areas of Zimbabwe saw famine creep in from usually vulnerable areas to the entirety of the population. The governmental response was to provide after-the-fact aid and, with this move, famine – or at least a 'famine that kills'[10] – became nearly non-existent in the country. Instead, the country's structurally poor suffered from chronic hunger and poor food entitlements across all regions. 'Food was regularly available to those who could afford it and regularly scarce for those who could not.'[11] This depiction of hunger in Zimbabwe illustrates how the imposition of a capitalist free market affected food security in the nation. Acute starvation was exchanged for widespread chronic hunger – attributed to the 'natural' fluctuations in the market economy. This meant, of course, that poorer citizens of Rhodesian society, composed mainly of native Rhodesians,[12] became food insecure. Dangarembga positions her novel within the context of this material history of deprivation. She paints a dismal portrait of everyday life in Rhodesia for its poorest inhabitants. In this capitalist-driven nutritional environment, social and economic mobility are the key to dietary security and freedom from the physically and mentally crushing experience of constant hunger. The novel indicates that money is the means to health, and education is the means to money.

The preliminary section of Dangarembga's novel establishes the overarching logic of food consumption and production that governs the characters' lives. Covered roughly in Chapters One through Four, this section presents the cultural, social, and economic structures that dictate the ebb and flow of daily life at the homestead and how these structures constitute individual characters' lives. Images of food and hunger permeate this section, as the connections between characters' attitude towards food, their roles in its production and consumption, and their daily mission to escape the clutches of food dearth are foregrounded as paramount to understanding life on the homestead. Narrated through the eyes of young Tambu, the economy of food production and consumption is so dominant on the homestead that even time is measured by its dynamics: 'The school year ended and the maize year began.'[13] That Tambu's family faces a constant state of food insecurity is made concrete from the opening of the novel. Although this atmosphere of deprivation is widespread, there is a particular focus on the dietary hardship of the female characters in the novel. This society is characterized by strict gender roles, and women become the inevitable objects of consumption *and* production.

In the novel, inequality is defined through both economics and gender. Female members take on the additional burden of womanhood and bear the brunt of the family's labour but reap the least reward in terms of nutrition. Tambu and her female counterparts oversee food preparation and land cultivation while the men of the family are portrayed as the indolent consumers of their labour. The women are forced to eat after the men have had their fill, and women receive a meagre and less varied diet in an already strained nutritional environment – meat is a privilege that is mostly enjoyed by men:

> the postcolonial hunger narrative ultimately showcases the alignment of power and foodways by asking us to consider not only who eats, how much, and in what order, but also whether the pleasures of food and eating are distributed equally, especially for women, immigrants, and other alimentary sub-citizens in the gastropolitical order.[14]

Women's hunger is experienced as a consequence of both colonially mediated hunger *and* patriarchy. Their bodies are subject to the hardships not only experienced by poorer Shona communities but also because of the patriarchal control that Dangarembga describes as endemic to Shona culture. Despite their essential role in the reproductive process of the family (reproductive in both senses – in reproducing the family's bloodlines *and* reproducing the family's subsistence by farming), women are viewed as an inevitable financial and social drain.

When Tambu's teacher, Mr Matimba, suggests to Tambu's father, Jeremiah, that Tambu's education has the potential to help lift his family out of poverty, Jeremiah answers: 'She will meet a young man and I will have lost everything.'[15] This contrasts with the family's willingness to educate her brother Nhamo. Sending Tambu to school is interpreted as a waste of resources. As is the case in Shona marital tradition, Tambu is expected to leave her father's home and move to her husband's family, taking the benefits of her labour with her. Her social worth is predetermined by the economic labour her body can produce – and as a woman this power is always limited and controlled by the man to whom it is culturally understood she must answer,[16] whether that is her father or her husband. This holds true even for Maiguru, Babamukuru's wife, who, despite occupying an economically privileged position in Rhodesia, nevertheless suffers the obligations of womanhood. This is reflected in the subtle politics of Babamukuru's westernized dinner table. 'Maiguru said that Babamukuru's old meal was no longer fresh. She said she would eat it herself, that Babamukuru should serve himself another portion of food [...] so the ritual dishing out of my uncle's food was performed again.'[17] Despite its veneration by the novel's narrator as the location of social enlightenment, the gender politics of Babamukuru's house mimics the patriarchal traditionalism of the homestead.

The first section of the novel chronicles Tambu's first step towards escaping the homestead, a place characterized by gender inequality, to the mission, which she constructs as offering her the means of overcoming it. Her brother and her father embody the agents of a patriarchal order from which she struggles to gain independence. The language and site of this struggle are articulated through food and eating. When Nhamo is accepted into the mission school and Babamukuru's household, he expresses his happiness by taunting Tambu: 'Pound well while I am eating potatoes at the mission!'[18] The measure of development, whether it be economic, social, or maturation, is expressed through food. In response, Tambu becomes determined to go to school herself, to acquire the same opportunities for social mobility, cultural freedom, and a better diet. She does this by cultivating a space of her own land to grow and sell mealies (maize) to raise the fees for school. Her attempts are discouraged not only by her brother and father but also by her mother: 'And do you think you are so different, so much better than the rest of us? Accept your lot and enjoy what you can of it. There is nothing else to be done.'[19] Her mother is demonstrative of the role of female complicity in the gendered economy of the homestead and expresses in a gentler way the ideologies her brother espouses cruelly:

'But I want to go to school.' [Tambu]
'Wanting won't help.'
'Why not?'
[Nhamo] hesitated, then shrugged. 'It's the same everywhere. Because you are a girl [...] That's what Baba said, remember?'[20]

Against the odds, Tambu grows mealies and even manages to sell them at a considerable profit to some sympathetic white settlers with the help of her former teacher Mr Matimba. With the money, she enrols at school, but her plans are almost foiled by her father when he claims the money as his own. He accuses the headmaster: 'That money belongs to me. Tambudzai is my daughter, is she not? So, isn't it my money?'[21] Jeremiah owns Tambu's physical labour, and as such she does not have rights to personal ownership or financial agency. Luckily, her father is stopped by the altruistic Mr Matimba, and Tambu re-enrols in school. However, the gendered economy of power remains intact – Tambu's father would not listen to Tambu but registers the authority of a male teacher.

Tambu's narrative makes obvious the gender divides that characterize the daily existence of the women that live in the homestead. Although her critique of the homestead is particularly inflected through gender taxonomies, another pervasive ideological system feeds into and influences homestead culture. The bodies described on the homestead must be read intersectionally, through the racial and gender politics of the colonialized female subject. Tambu has little direct interaction with the colonial power in the country, but it nonetheless exerts a powerful influence on and throughout the narrative. The intersecting positions of colonial subject and female are explored in *Nervous Conditions* as complicit and conflicting. Development narratives of Modernity espoused by colonizers are exacerbated by Shona patriarchal practices, producing a kind of compounded 'nervous condition' in the native female population. The Shona female body is the overdetermined site of both colonial and patriarchal oppression, and the material realities of this inequality are expressed on the female form. Although the book's intervention into the gender issues that characterize Rhodesia's socio-economic structure stands out as the immediate and most obvious focus, it also explores how this system of inequality fits into, interacts with, and is informed by colonialism and race. Dangarembga explores the effects of colonial rule in less overt ways, but they are also expressed through the language of food and hunger.

The representation of colonial rule is mediated via white characters, but there are very few examples in the novel. This speaks not only to the informal and formal racial segregation that separated the white population from the native, but also makes clear how wealth is distributed along racial lines. The narrative's predominantly invisible colonial presence indicates its preoccupation with the psychological and ideological traumas of the native population, or their 'nervous conditions.' However, there are ways to read the pernicious effects of colonialism in the novel, although they are mostly mediated by interactions between native Rhodesians. Mr Matimba demonstrates an awareness of colonial power and its racist strategies in the segregated white-minority-led nation. When he suggests that Tambu might earn greater profits if she sold her mealies to white Rhodesians, his suggestion emerges from an understanding of how power is constructed and reinforced in the colonized nation. His strategy to help Tambu comes with the experiential knowledge that the product is not as important as the sale when interacting with the white population of Rhodesia. The obsequious way he offers the mealies – 'sorrowfully and beseechingly'[22] – to white shoppers in Umtali demonstrates his cognizance that interactions with colonial forces require a presupposed submissiveness. He must perform the familiar role of the needy native, desperate for charity. He knows that to ensure a sale, a sympathetic appeal to the

Christian values of white Rhodesians is the most effective. Mr Matimba goes on to exaggerate Tambu's poverty and social circumstances to achieve the best sale, and after a good dose of colonial guilt, the mealies sell at the relatively astronomical sum of ten pounds. Mr Matimba is commended for his part in Tambu's envisaged struggle out of poverty; Doris (the elderly lady who purchases the mealies and is declared 'more human than most of her kind') departs with a sense of Christian charity, and Tambu is deemed a 'plucky piccannin.'[23]

Mr Matimba's interaction with Doris is a rare exchange in *Nervous Conditions*, a novel in which white Rhodesians appear sporadically; and when they do, they are essentialized in a manner that does not fully represent the complex, subtle, and often powerful ways the individual colonialist–native interaction impacts Rhodesian culture on a macro level. Instead, the narrative focuses on the native population and explores how imperial discourse shapes the polarized social environments of the homestead and the mission – both spaces inhabited predominantly by black Rhodesians. Dangarembga's construction of the homestead, and the various daily hardships of its inhabitants, belies a more insidious history of hunger at play in both Rhodesia's past and present. Dietary norms on the homestead are constructed within the parameters of an ever-present hunger, and this is normalized by the narrative as a means of raising the reader's awareness of a pervasively low nutritional standard – but one that is so prevalent that it is matter-of-fact. The continual threat of death from starvation indicates the dietary norm on the homestead is that of severe malnutrition – a chronic situation that constantly teeters on the brink of acute starvation: for Tambu, 'famine is rooted in the normal, in the prosaic, and in the everyday. [...] Famine contains the terror of the possible.'[24] Here, the precariousness of life and the frightening proximity to starvation are recognized as simultaneously mundane and agonizing. 'Dangarembga encourages us to consider health and illness in situated cultural terms.'[25]

Tambu's narration incorporates this 'terror of the norm' by almost dismissing it. It is only casually included in the text: 'Although we harvested enough maize to keep us from starving, there was nothing left over to sell.'[26] The potential consequences are repeated throughout the text in this mundane tone: the rhetorical tactic further emphasizes the proximity of severe hunger. As Tambu's father and brother prepare to take a train journey of a few hours, Tambu's mother frets over their food provisions for the journey, stating that 'they should not blame her if they slept at my aunt's homestead only to die of hunger on the train.'[27] The intended tone of the statement is unclear given the context of hunger in Rhodesia; the reader is made to feel unsure whether the statement accurately reflects the reality facing these poor characters or whether it is a turn of phrase. The rhetorical ambiguity brings attention to the normative status of extreme food insecurity present in the text while highlighting a disparity between it and a western audience's expectations. '[T]he interplay between westernized and indigenous Shona cultural formations and responses to food destabilizes any such concept of a "normal" state of health.'[28] Awareness is drawn to the extremity of the novel's foodscape, but the text itself remains silent on the matter of causality. The onus is placed on the reader to decipher the material connections.

In the scene where Doris purchases Tambu's mealies and 'saves' Tambu from a future without education, Mr Matimba presents both himself and Tambu through deploying the trope of the starving African – an object of sympathy. But he is careful to not present his case as demanding or seeking reparation. Tambu, Mr Matimba, and Doris are complicit in a system of exchange whereby money changes hands, but doing so only further entrenches tacit power inequities, and these inequalities codify food as objects that signify power and access to power, and reinforce racial hierarchies. In the same episode, Mr Matimba indicates his own self-awareness and collusion in the colonial dynamic, when he and Tambu discuss Umtali's (the city they reside in) infrastructure:

> The road began to climb upwards on the shoulder of the hill. The truck faltered, and changed its voice and moved more slowly.
> 'The white people must be very strong to build such a wide road so high up,' I observed.
> Mr Matimba did not think so. 'We did the building,' he told me. 'It was a terrible job. We did many terrible jobs. Now we are approaching the top of Christmas Pass,' he said, changing the subject.[29]

Mr Matimba's swift verbal diversion at the end of the exchange indicates his understanding of the systems of power that operate within the nation, and a possible anxiety regarding his role – collusive or coercive – in upholding its structures. His description of the road reveals an understanding that the weight of colonial rule is usually borne by the colonized, and acknowledges the role of the colonial body, and its labour, in the colonial system. The novel points to a material history of the native body, not limiting its exploration of the body within discourse, or its symbolic powers. Mr Matimba's reluctance to criticize the injustice of colonial rule repeats the novel's tendency to avoid directly commenting on imperialism. Instead, the impact of colonialism is embodied by Tambu's cousin Nyasha. Dangarembga posits a link between the native body and the 'weight' of colonialism, and then explores how this burden is compounded by gender inequalities and is expressed through the weight issues of her female characters. The intersection of gender and race produces the doubly burdened female bodies in the novel, and the focus is placed on Nyasha and her disordered eating.

Weighty Issues: Female Bodies in Nervous Conditions

Despite the narrator's dichotomizing of the homestead and the mission, the novel uniformly constructs all of its female Shona characters as the primary bearers of hardship – and consequently, hunger – regardless of their social class. The novel's representation of female aesthetics demonstrates that female identity is comprehended through a double bind of colonial subject *and* womanhood, irrespective of the woman's socio-economic status or the position of the men they are married or related to. Tambu's mother succinctly states this problem: 'As if it is ever easy. And these days it is worse, with the poverty of blackness on one side and the weight of womanhood on the other.'[30] In her article 'Disembodying the Corpus,' Deepika Bahri makes clear the connections and collusions between patriarchy and colonialism. She states:

> By layering gender politics with the atrophying discourse of colonialism, Dangarembga obliges us to recognize that the power structure is a contradictory amalgam of complicity and helplessness – where colonizer *and* colonized, men *and* women collude to produce their psycho-pathological, in a word, 'nervous conditions.'[31]

In response to the dual pressures of femininity and colonial racial identity, the ideal of Shona beauty is radically different from the western standard. The female characters of the homestead privilege a rounder, healthier female body. Tambu's grandmother describes this ideal female body:

> At one time I was as small and pretty and plump as you, and when I grew into a woman I was a fine woman with hair so long you could plait it into a single row down the middle of my head. I had heavy, strong hips.[32]

As Clare Barker notes: 'This indigenous construction of beauty in terms of health promotes fatness as a desirable state, gesturing toward historical conditions of hardship caused, in part, by colonialism, and containing within its meaning the attractiveness of an ability to work the land.'[33] Like Bahri, Barker acknowledges the interplay of gender and colonialism that claims women's bodies as the site of its expressions. These Shona women embody a context of food insecurity, and a legacy of historical hunger – which also situates the female body in the nation's colonial past and present. The fat of the idealized Shona body insulates within it a history of Rhodesia's legacy of starvation and dietary hardship. Female bodies are constructed as the sign-bearing site of colonial exploitation and patriarchal subjugation. Nyasha's food abnegation is a response to these burdensome historical significations as they re-form and re-enact in Rhodesia's present socio-economic environment. As such, Nyasha's hunger strike can be read ambivalently. It can be interpreted as a denial of colonial history and its reverberations, because 'To starve is to renounce the past [...] because it is to void the body of its stored anteriority.'[34] Yet in the startling image of her starvation, Nyasha also re-invokes a suppressed colonial history of deprivation and hunger and can be read as a project of historical reclamation, as well as a strategy to reclaim the black female body.

Critical work on *Nervous Conditions* has largely focused on the issue of Nyasha's problematic relationship with food: this arises largely due to the purportedly unusual nature of her hunger. 'The largely acclaimed *Nervous Conditions* by Zimbabwean author Tsitsi Dangarembga has generated countless articles due, of course, to its literary quality but also to its treatment of a pathology not usually associated with Africa.'[35] Given Rhodesia's prevalent food insecurity, a voluntary food abnegation is read as a misnomer. In combination with this contradictory historical context, readings have often foregrounded the notion that anorexia is a specifically western phenomenon with little representation in Africa.[36] Critics have consequently read her as an example of a colonial elite who neurotically mimics the culture of her oppressors, claiming that 'Nyasha's anorexic body is a parody of a Western style ideal of slim, feminine sexual desirability.'[37] There is certainly evidence within the novel to support these readings, and Nyasha's food refusal can and should be read within the context of these theories of western pathology. Nyasha's English upbringing necessitates this interpretation:

> With thirteen extra people to feed – and the lot of us devouring seven loaves of bread and half a pound of margarine each morning, to say nothing of sugar, because (with the notable exception of Nyasha, who believed that angles were more attractive than curves) we liked our tea syrupy.[38]

Nyasha's hybrid cultural and national identity is certainly a factor in her self-imposed hunger, but the presumption that this is the only cause of her anorexia is problematic. It subsumes Nyasha's body into a pathologized western discourse and immobilizes her body as a static signifier – a passive container of an imitation of the idealized white, western female body. This kind of reading undermines the possibility of reading Nyasha's body as a productive site of articulation, readings that, I argue, the novel's form, context, and setting themselves suggest. Alongside these anorexia-as-pathology readings, Nyasha's starving body can also be read as a reassertion of the national history of colonial deprivation. It can signify as a rejoinder to the racist dualities that structure the civilizer/civilized ideologies of imperialism. It can also be read as a strategy of resistance against the twinned biopolitical technologies asserted by colonialism and patriarchy. These potential meanings are omitted if Nyasha's body is only read within the western context of anorexia. To dismiss Nyasha as a circumstantial anomaly is reductive: 'If Nyasha is something of a curiosity in an African context, in what sense is it possible to speak meaningfully of her "anorexia" or "bulimia"?'[39] I contend that Nyasha should be read not in spite of the unusual context of her hunger, but particularly *because* of them. This reading facilitates a contextually sensitive understanding of why Nyasha's protest takes the particular form of hunger that it does, and also goes some way in comprehending why, as Tambu's narration intimates, it 'may not have in the end been successful.'[40] As with all the hunger strikes examined in this book, Nyasha's protest produces ambivalent meanings and contains ambiguous political outcomes.

Nyasha's illness can be read as responding to the conflicts and dislocations that arise from the hybridity of her identity. Several theorists have written along these lines and have figured Nyasha's eating disorders as a means to combat the double threat of patriarchal and colonial subjugation. Bahri states: 'Nyasha's war with the patriarchal and colonial systems is fought on the turf of her own body, both because it is the scene of enactment of these systems and because it is the only site of resistance available.'[41] In the same article, Bahri gestures towards a more optimistic reading of Nyasha's dwindling body and state of health:

> Nyasha's offensive against her bodily self reenacts the narrative of violence on woman and native while at the same time gesturing at the possibility of agency: signaling from the bathroom and the bedroom (her favorite retreats) that a more pervasive insurgence, a more public and widespread struggle by women for freedom from the patriarchal and colonial order may soon be to follow.[42]

In her article 'The Nervous Collusions of Nation and Gender,' Heather Zwicker responds to this summation and reading of agency by signalling her own reservations, noting:

> This figurative reading of Nyasha's eating disorder is intuitively compelling to a certain degree [...] but there is a limit to the usefulness of such a reading [...] although this reading accords the anorexic *some* agency, it seems to curtail what that agency might look like (i.e., it's always self-destructive) from the outset.[43]

Zwicker is not alone in indicating the problems associated with imagining the eating disorder as a productive site of resistance. Nyasha's dissent is certainly a self-destructive one, and the consequence is bodily harm, delirium, and a psychiatric evaluation.

For Bahri, the symbolic agency that Nyasha's hunger represents is enough to qualify it as politically useful, but Zwicker finds this possibility somewhat debilitating, as it does not provide any real-world examples of anti-colonial activism, outside of self-styled bodily violence. These debates replicate the problematic of Michael K's hunger explored in the previous chapter. Susan Z. Andrade states that 'although Nyasha's story produces an ending without a positive role model, her outburst here suggests that she, and through her, the reader, has arrived at such a moment of productive crisis.'[44] The novel concludes at this frontier of possibility. I examine this moment of productive crisis through reading Nyasha's psychological and physical transformation in conjunction with Tambu's narration. The novel offers hope for female agency, while still problematizing the somatic and discursive violence of the hunger strike. Thus, hunger and the novel's conclusion produce ambivalent readings.

Starving Anomaly: Nyasha's Hunger Strike and Nervous Condition

There is a wealth of critical work on eating disorders.[45] This scholarship goes beyond nutritional and medical frameworks to explore self-starvation's socio-historical meanings. Based in the humanities, social sciences, and psychology, this work posits the body as a discursively constructed site of meaning. The theory of the body that is most relevant to this study is the post-Enlightenment Cartesian body, and I read anorexia as an attempt to intervene into its binary logic. Susan Bordo deploys this model of subjectivity in *Unbearable Weight* and pays close attention to the gendering of the Cartesian body: The passive, material, emotional, and uncontrollable body is coded as female, while the active, immaterial, rational, and controlled mind is coded as male. In this distinction, women are associated with sensuousness – sexual instinct and animal appetites – and men are associated with the more civilized pursuits of philosophy and the mind: 'the inevitable, like a pure idea, like One, the All, and the Absolute Spirit.'[46] Bordo describes various strategies that subjects use to situate themselves and others in the relational logic of the dualism: both female and male bodies internalize this gendered binary, emphasizing their complicit engagement with its forms. The female body is the target of the disciplinary logic that this dualism produces. In this vein, Bordo offers the most widely accepted explanatory model for eating disorders:

> Anorexia Nervosa [...] can be seen as at least in part a defence against the 'femaleness' of the body and a punishment of its desires. Those desires have frequently been culturally represented through the metaphor of female appetite. The extremes to which the anorectic takes the denial of appetite (that is, to the point of starvation) suggests the dualistic nature of her construction of reality: either she transcends the body totally, becoming pure 'male' will, *or* she capitulates utterly to the degraded female body and its disgusting hungers. She sees no other possibilities, no middle ground.[47]

The anorectic's struggle, realized in these terms, is not merely a reaction to the unreasonable pressures of aesthetic ideals. It is an attempt to literally transform the body and consequently to reconfigure subjectivity. Therefore, it can be read as a response to the ideals of femininity that so often play out on the female body:

> Whenever women's spirit has been threatened, she has taken control over her body as an avenue of self-expression. The anorectic refusal of food is only the latest in a series of women's attempts at self-assertion which at some point have descended directly upon her body. If a woman's body is the site of her protest, then equally the body is the ground on which the attempt for control is fought.[48]

The anorectic's agenda is to discipline and contain the femaleness of the body, to bring under control not only its shape but also the associated qualities – of sensuousness, a lack of control, sexual excess, and hysterical emotion. 'Anorexia is, above all, an illness of self-division and can only be understood through this tragic splitting of body from mind.'[49] It is no coincidence that part of the anorectic's strategy is to achieve an emaciated figure – far above and beyond what is considered 'beautiful' in the western context. Through extreme food refusal, the anorexic evacuates the curves and obvious physical indicators of femininity from the body. The starving body transforms into a genderless form. Hunger thus presents an ambivalent agenda as resistance: 'how exactly is the text of contestation articulated in the anorexic female body when the latter is frozen in her preadolescent stasis and refuses to become female?'[50] Theories of anorexia converge on socially constructed ideas of femininity. They read the female body as both a site of agency and protest: 'Anorexia is a political system, a micro-politics: to escape from the norm of consumption in order not to be an object of consumption oneself,'[51] but also as the restrictive social codes of femininity taken too far: 'the logical conclusion of the impossible and contradictory messages society sends her.'[52]

Nyasha's self-imposed hunger can be read as an intervention not merely of gender but also of the racialized colonial body. Her anorexia protests against these competing discourses as they play out their significations on her body. Her hunger is mostly directed at her father – a product of the system of colonial power who as such represents both patriarchy *and* colonial power. The gendering of the colonial female body is especially relevant when considering its function within the discourses of imperialism. 'Many forms of colonial writing and art produced by Europeans used the figure of the woman's body to suggest that nature formed the physical and mental landscape of the colonized.'[53] The essentialized female body is localized as the primary site of imperialism's civilizing process. Thus, Nyasha's eating disorder can be read as responding not only to a dominating patriarchal order but also to a system of inequality whose parameters are defined by a colonial history and culture. This manifests in the incongruence of wealth and power made manifest within the larger family unit. If Nyasha's hunger is intelligible only within the context of all these competing ideologies and tensions, then it may be said that her anorexia/bulimia is a 'struggle to release the body from all contexts, even from the context of embodiment itself.'[54]

Tambu underscores the body/mind duality in her narration. After Nhamo's death, Tambu moves to live with her cousin and uncle at the mission to go to school. Describing her initial sense of arriving at the mission, she states:

> At Babamukuru's I would have the leisure, be encouraged to consider questions that had to do with the survival of the spirit, the creation of consciousness, rather than mere sustenance of the body. This new me would not be enervated by smoky kitchens that left eyes smarting and chests permanently bronchitic.[55]

Here, Tambu overtly privileges the mind over body; moreover, she references the kitchen – a domestic space that is coded as a site of female entrapment. Her imagined emancipation from the material realities of her female body is enabled by access to a better education. Books, discourse, words – technologies of the mind – are privileged over the domestic technologies of the body – producing food to feed other bodies. Education means freeing her mind, which will also free her body from the chains of its material obligations; in her case, the social and financial realities of being a woman born to a poor Rhodesian family. Like Tambu, Nyasha's aspirations obey a similar logic, although she eventually takes this logic to an extreme through her disordered eating.

Nyasha's food refusal occurs early in the text, during her initial visit to the homestead after her return from England. It is here that Tambu first notices her nascent tendency towards food abnegation. The event confuses everyone, particularly Tambu, who interprets Nyasha's refusal as rude. The way she refuses food provides clues that portend something beyond the usual diagnosis of anorexia.

> I missed the bold, ebullient companion I had had who had gone to England but not returned from there. Yet each time she came I could see that she had grown a little duller and dimmer, the expression in her eyes a little more complex, as though she were directing more and more of her energy inwards to commune with herself about issues that she alone had seen.[56]

This description recounts the turn toward interiority that accompanies Nyasha's food refusal; she seeks nourishment for the mind rather than the body. As she inwardly wrestles with these private issues of the mind, the reader is directed to the possible source of Nyasha's meditations: 'Nyasha, who had an egalitarian nature and had taken seriously the lessons about oppression and discrimination that she had learnt first-hand in England.'[57] The connection between reading and eating is vaguely established in this initial instance, but is made clearer in Nyasha's first confrontation with her parents at the dinner table at the mission:

> 'I don't mind going to bed hungry,' said Nyasha.
> 'When did you ever go to bed hungry? Not in this house!' snapped Maiguru.
> 'All I wanted to say,' Nyasha replied apologetically, 'was that when I can't sleep usually what I need is a good read. Really! Sometimes I have to read until one o'clock, but after that I usually drop off.'[58]

Nyasha's replacement of words for food – and her tacit insistence that 'food' for the mind is enough to sustain the body – reinforces the binary logic of the Cartesian duality. The idea of words taking the place of food differentiates Nyasha from Michael K and Nimi from previous chapters. Their hunger is accompanied by silence. Instead, Nyasha substitutes food for words.[59]

The relationship between food and words is long-established and can be traced back to at least the Old Testament: 'man doth not live by bread only, but by every [word] that proceedeth out of the mouth of the LORD doth man live.'[60] This Old Testament verse materialized in the performance of rituals that make clear the link between bodies and divine affirmation. 'Ingesting the foods "inscribed" with the word of God is a ritualization of scriptural metaphors.'[61] Ritual as metaphor was taken to more literal ends in the Early Modern period, when 'to reflect upon the materiality of language was to experience words as edible.'⁶² Although this may seem alien to us now, Early Modernists believed in the sustaining quality of words on a physiological level. The metaphor of consuming knowledge through food and feasting on words has purchase in the construction of the Modern body, although in more symbolic ways: 'The mouth is [...] a potent symbol of both consumption and its control, combining in the one site sensuality/nature (the tongue and the tastebuds) with rationality/culture (the organ of speech).'⁶³ The ingestion of both foods and words position the body, and its oral boundaries, as a site of alimentary and social transformation. To eat and starve is to alter the self, and to read/speak/write or to remain silent mimics this alimentary logic. Ellman states that the 'vampiric relationship of words to flesh typifies the literature of self-starvation.'⁶⁴ In this formulation, the body is nourished by words, and food is substituted by its discursive power.

Following the exchange at the dining table, Nyasha is caught reading D.H. Lawrence's *Lady Chatterley's Lover*. The book is deemed inappropriate by Babamukuru; its transgressive narrative of female sexuality as well as the western literary tradition from which it emerges is significant. Nyasha seeks sustenance in sources of knowledge that defy her father's traditionalism. This strategy is complicated by the fact that her chosen text is from imperial literary tradition. If Nyasha tries to 'starve out' the significations of her colonially inscribed black body using the freedoms proffered by western feminism, she is again faced with imperialism's civilizing impulses. Nyasha's split subjectivity (an identity whose materiality contains the violence of colonialism) conflicts with the narratives it ingests: '"I don't know what's wrong with our daughter. She has no sense of decency whatsoever," says Babamukuru, requisitioning the book from Nyasha, "[...] No daughter of mine is going to read such books."'⁶⁵ In this scene, we are reminded that Nyasha is subject to similar modes of objectification to Tambu, despite Nyasha's education and socio-economic advantages. Her body is mediated by a patriarchal control that monitors the materials she puts into it – whether they be food or words.

Nyasha's initial food refusal develops into more severe forms as the narrative progresses. During Tambu's stay at the mission, Nyasha denies her body food and crams her mind with words in its stead:

> She was working much harder than I had ever seen her work before, up long before her usual time, so that when breakfast was ready she had been studying in a concentrated state for an hour or more. At night it was the same: by eight o'clock she was curled up in bed with her books, but the light rarely went out before one. Everybody agreed that she was overdoing it. She was looking drawn and had lost so much of her appetite that it showed all over her body in the way bones crept to the surface, but she did not seem to notice.⁶⁶

Nyasha becomes anxious, desperate in her appetite to consume books and knowledge. 'As if there's everything to learn and I'll never know it all. So I have to keep reading and memorizing, reading and memorizing all the time.

To make sure I get it all in.'[67] Nyasha feeds her mind to the detriment of her body. By transferring her identification with the black female colonial body to the masculinized mind, she retreats to a space of possibility, and freedom may be imagined. The material body is counter-constructed against the intellectual/spiritual category of the mind that needs only words for nourishment. Nyasha's ravenousness for knowledge exposes the implications for her subjectivity – her panic to make sure she 'gets it all in' defines her inner psychic self as separate from her somatic self, with the boundary becoming progressively more difficult to sustain.

Nyasha's subject is under siege. Her strategy to overcome the racialized, female, colonial restraints of her body is enacted by starving it of nourishment and instead feeding the male-centric, 'pure' spirit of her mind. But as she descends further into this negation of body and exaltation of mind, the interconnectedness between the two is reinforced: as her body starves, it deprives her mind of its cogency. Thus, the structure of the protest itself is based on faulty logic – constructed on the grounds of an inextricable oppositional binary. The body cannot be escaped, literally and linguistically. The mind does not, and cannot, exist without it. Reading Nyasha's Cartesian strategy deconstructively produces wider implications for the ingestion of discourse, and, more significantly, western discourses and literature. The words that Nyasha ingests provide no nourishment; they do not represent her colonial, female body. The feminism that might be found within this writing does not account for her lived experience – rather she is marginalized in its progressive ideals. Nyasha's hunger strike contains and repeats the contradiction, and self-division, of the colonial subject – doubly so given her position as a young Rhodesian woman. She attempts to disembody herself, but the resources at her disposal prove more harmful than liberatory. She cannot find the language to write an alternative narrative for her colonial, female body.

In the final chapters of the novel, we witness Nyasha's eating disorder crescendo. Contrasted with the initial dinner scene where Nyasha's polite refusal of her meal is registered only by Tambu, the final dinner scene turns violent, mirroring the acceleration of Nyasha's bodily self-harm. As Nyasha refuses to eat yet again, Babamukuru enforces his authority to try to stop her transgressions and attempts to force Nyasha to eat: 'She is always doing this, challenging me. I am her father. If she doesn't want to do what I say, I shall stop providing for her – fees, clothes, food, everything.'[68] In response, Nyasha docilely consumes all of her dinner, and then promptly throws it all up in the bathroom. This scene is particularly significant, as Nyasha chooses this point to reveal the reasoning behind her food refusal in no uncertain terms:

> Imagine all that fuss over a plateful of food. But it's more than that really, more than just food. That's how it comes out, but really it's all the things about boys and men and being decent and indecent and good and bad.[69]

Nyasha communicates her own reading of her father's food as signifying the multiple constituents of a patriarchal, colonial, social, and economic order.

With her starving body, she protests the multiple forces that work together to produce the food on her father's table: his complicity with the colonial mission, his enforcement of unequal gender practices, and even the class and cultural snobbery he displays in his food choices. She dissents in the same arena that these forces encode their powers: her body. Ellmann traces the desire for disembodiment contained within the will of the hunger striker. It is an attempted evacuation of the history of suffering written on the starving body. 'The starving body is itself a text, the living dossier of its discontents, for the injustices of power are encoded in the savage hieroglyphs of its sufferings.'[70] Nysha tries to obliterate this body, along with its narrative of oppression. Nyasha's hunger strike is a paradoxical response to a paradoxical subject position, but in many ways it is the most logical action for her to take. Read in this way, her protest is not successful in so far as it results in her own suffering and bodily harm, but her intervention is available for us to read the contradictions of the female colonial body, and identity. Thus, the text of Nyasha's suffering is more productively read than it is lived.

In the final chapter of the book, Nyasha's body and mind are both crippled from the regimen she uses to 'discipline [her] body and occupy [her] mind.'[71] She mentally deteriorates as she struggles to retain information, and experiences lapses in concentration and memory. On the eve of Tambu's transfer to the Sacred Heart mission school, Nyasha reaches the culmination of her 'horribly weird and sinister drama.'[72] She rants incoherently, raging against the multiple forces of domination inscribed on her body for which she has sacrificed both her mental and physical health to defy:

> 'Why do they do it, Tambu,' she hissed bitterly, her face contorting with rage, 'to me and to you and to him? Do you see what they've done? They've taken us away. Lucia. Takesure. All of us. They've deprived you of you, him of him, ourselves of each other. We're groveling. Lucia for a job, Jeremiah for money. Daddy grovels to them. We grovel to him.'[73]

The referents in the outburst are difficult to pin down – revealingly so. The potential misrecognition in the pronouns betrays the complicit and complicated history of colonialism that produces subjects like Babamukuru, his extended family and of course like Nyasha: 'I'm not one of them but I'm not one of you.'[74] Nyasha communicates through a rhetoric of divisiveness: the split subjectivity that characterizes her hybrid identity, the socio-economic differences that divide the homestead and the mission, and the racist segregation of Rhodesia. Along with her verbal attack, Nyasha ravenously attacks her books, tearing at them with her teeth in a figurative engorging: 'Their history. Fucking liars. Their bloody lies.'[75] Nyasha's attempt to retreat into the realm of the mind can be interpreted as a move away from the material, colonial contexts of her body. But the western discourses she seeks answers from only present her again with a model of abject subjectivity as a colonial subject. Instead of consuming books in lieu of food, in this scene she rejects all forms of nourishment – mental and physical. Her body and mind are defeated – in attempting to nourish one to the exclusion of the other, the dependent formation that characterizes her body and mind is only further exacerbated. This, in fact, demonstrates that Nyasha's self-division is an inherited colonial construction; in Fannonian terms, she is filled with the self-loathing contained within colonial discourses and the education that she has ingested, as a hybrid subject. Her hunger strike is a bodily manifestation of this self-division, and the Cartesian model of the body does not provide her desired enlightenment. She cannot negate her body or exist outside of it.

By taking Rhodesia's history of famine into account, Nyasha's body can also be read as aligning itself with the starving bodies of her famished countrymen and their history of nutritional hardship. As Richard Gordon states: 'Many anorexics in fact identify explicitly with the "wretched of the earth," and their food refusal becomes a token of rebellion against the values and attitudes of middle class affluence.'[76] Read through this, Nyasha's rebellion responds ambivalently to the history of her starving body. Her performance of hunger makes visible the history of the food-insecure Rhodesian body, and the colonial factors that influenced it. However, in her attempted denial of the material histories written on her body (by 'starving it out'), she also re-enacts a symbolic denial of these histories of colonialism, because the materiality of her body is allied with the material history of Rhodesia. Nyasha's hunger can be read as a productive symbolic protest against the competing oppressions of race and gender as they are contained in her body, but again this agency is curtailed not only by the inevitable physical consequences of her starvation but also by the wider question of whether this kind of somatic practice can be read in a politically productive way. '[Anorectic] [w]omen get ill instead of organized.'[77] Anorexia can be read as a force of will, a display of self-control, but the starving female body is also weakened and infantilized through this rationalizing somatic technique.

Following her explosive episode, she confides in Tambu:

> The next morning she was calm, but she assured me it was an illusion, the eye of a storm. 'There's a whole lot more,' she said. 'I've tried to keep it in but it's powerful. It ought to be. There's nearly a century of it,' she added, with a shadow of her wry grin. 'But I'm afraid,' she told me apologetically. 'It upsets people. So I need to go somewhere where it's safe. You know what I mean? Somewhere people won't mind.'[78]

Nyasha tacitly refers to the novel's situation within colonial histories: 'There is nearly a century of it.' Her efforts to 'keep it all in' also indicate that she constructs her body according to the idealized 'closed' post-Enlightenment body, policing the boundaries of her body and the materials that enter it. However, this proves impossible; her body and identity are read in the context of its relational structures. Her body is not closed, but permeable. The self-disciplinary bio-practices of the idealized, individuated body are attempted, but fail, just as the attempted wresting open of the divisions of body/mind fails. Her attempts to make sense of her body and its place within society cannot be divorced from the heavy burden of her body's history. Her fractured identity is a reality she cannot deny, and it cannot be starved into submission – creating a unified, clearly bounded subject.

Critics have noted the ambivalences of Nyasha's hunger.[79] Nyasha's rebellion concludes in a psychological break with reality, and although this can hardly be read as successful or productive, it does represent an attempt at agency – even on a symbolic or semiotic level:

> Even though the anorectic body seems to represent a radical negation of the other, it still depends upon the other as spectator in order to be read as representative of anything at all. Thus its emaciation, which seems to indicate a violent rebuff, also bespeaks a strange adventure in seduction.[80]

This situates the body within a larger system of discourse from which it cannot be extracted, where bodies and their meanings are inserted into a system of dynamic relations, shifting and contingent. It reinforces the constitutive discursive systems of the hunger strike. The starving body interrogates the relationship between self and other through its bodily technique – and, as such, can affect the bodies/identities of others by troubling the border between self and other. This speaks to the transformational power of the body. Nyasha's extreme disciplinary and discursive techniques of hunger fail in that they do not allow her to escape the somatic. As Bordo intimates, the anorectic swings from one extreme to another: 'She sees no other possibilities, no middle ground.'[81] Although the tactics of the anorectic read as an attempted inversion of the body/mind (and thus colonizer/colonized) hierarchy – to be the masculine mind rather than the feminized body – the failure to invert the terms only entrench the binaries further. To read Nyasha's hunger deconstructively is to lay bare the binary logic of the closed, Cartesian body. By doing this, we can read the discursive category of the body as an affective force, and the hunger strike as productive – even if it is not quite how Nyasha may have intended. The body may be read as revealing something of the truths of colonial discourse, power, and the body. The intended speech forms of Nyasha's hunger may be overshadowed by her physical deterioration, but we can choose to read the text of her body in a variety of ways. Her somatic speech is read and interpreted by Tambu to productive ends.

Hunger Denied: Tambu's Vampiric Narrative

It is easy to forget that Nyasha is not the main protagonist of *Nervous Conditions*, as critical work on the book has overwhelmingly focused on her hunger. However, the first-person narrative reminds us that the text is entirely focalized through Tambu. 'I was not sorry when my brother died.'[82] This controversial opening reminds us that there is a confident and defiant narrator at the heart of this story. Written in the retrospective mode of the matured subject and already having gone through the Künstlerroman form, Tambu is a wise and knowing narrator. She states: 'Something in my mind began to assert itself, to question things and refuse to be brainwashed, bringing me to this time when I can set down this story.'[83] The Tambu who narrates the novel exists in two times at once – she is split between the naïve Tambu who transforms in the present of the novel, and the experienced

Tambu who narrates this journey after the fact. The combination of the two simultaneous identities creates a sort of doubling in the narration. The distance created between the two narrative voices structures the didactic tone of Tambu's narration:

> It was up to them [Tambu's immediate family] to learn the important lesson that circumstances were not immutable, no burden so binding that it could not be dropped. The honor for teaching them this emancipating lesson was mine. I claimed it all, for here I was, living proof of the moral. There was no doubt in my mind that this was the case.[84]

These words are spoken at the start of Chapter Four, as Tambu leaves the homestead for the first time to live with her uncle at the mission. Although readers are yet unaware of the disillusionment Tambu experiences because of her experiences at the mission, the obvious moralistic, almost self-important, narrative tone foreshadows it. The 13-year-old girl who believes in the emancipatory potential of education constructs a stark ideological difference between the poor homestead and the idealized mission. By witnessing the deterioration of Nyasha, the tyrant-like behaviour of her uncle, and the oppression of her aunt, Tambu goes from naivety to knowledge. The 'limitless horizons'[85] Tambu initially envisages are, toward the end of the novel, considered through the complex racial, gender, and economic inequalities of Rhodesia.

However, it is rarely Tambu who undergoes the oppressions of biopolitical and physical control in the novel. They are instead displaced upon the other female bodies in the novel – upon Nyasha in particular, but also Tambu's mother and her aunt Maiguru. Little attention is given to Tambu's bodily development. This is because Tambu's body is underrepresented in the text. Although we understand that her body is subjected to the same social standards that the other Rhodesian women experience, the dramatic 'rebellion' in the novel is played out on Nyasha's body. Even her mother's body experiences violence in the form of illness and anxiety – again, the cause is gender inequality:

> In the week before I left she [Tambu's mother] ate hardly anything, not for lack of trying, and when she was able to swallow something it lay heavy in her stomach. By the time I left she was so haggard and gaunt she could hardly walk to the fields, let alone work in them.[86]

Tambu's aunt, Maiguru, is another objectified woman in the novel. Despite having as much education as her husband, not only is Maiguru responsible for the domestic duties for her own family but these duties multiply when she visits the homestead. Maiguru treats Babamukuru sycophantically and always performs her role of dutiful mother, but even she reaches her limit at one point in the novel and walks out on her husband for five days. Although this is a relatively minor example of resistance in the novel, these small acts of rebellion can be found – enacted by everyone except Tambu.

As has been previously discussed, Nyasha utilizes her own body to articulate her rebellion against the dual subjectivities of womanhood and colonial subject, and deploys a Cartesian logic to do so. Tambu subscribes to this body/mind model, as has been already explored in this chapter. Although it is Tambu who originally makes obvious her aspirations to 'consider questions that had to do with survival of the spirit, the creation of consciousness, rather than mere sustenance of the body,'[87] it is her cousin Nyasha who makes this idea literal in her hunger/anorexia. Although Tambu's narrative is sympathetic toward her cousin's protest and clearly approves of the underlying anti-colonial, intersectional feminist discourse that structures it, she remains a passive body in the text. Instead, she exists at the limits of food abnegation. The elision of Tambu's body is notable in the text. Instead, the responsibility of interrogating the female body is deferred to other women in the novel.

Tambu initially condemns Nyasha's behaviour, but as the novel progresses, this attitude softens to respect. At the start of the novel, Tambu claims that 'Babamukuru was right! His daughter was beyond redemption!'[88] But indignant censure gradually transforms into grudging admiration, until the narration makes clear Tambu's desire to follow Nyasha's example:

> Nyasha gave me the impression of moving and striving towards some state that she had seen and accepted a long time ago. Apprehensive as I was, vague as I was about the nature of her destination, I wanted to go with her.[89]

To transgress against the god-like figure of her uncle, however, is a step too far. 'I wondered: if I grew more used to my uncle, would I stop deferring to him, as Nyasha had, and I banished the thought from my mind because it was so dreadful.'[90] Babamukuru never has cause to discipline Tambu, as she disciplines herself. She elevates Babamukuru's rules to the point of the divine: 'surely it was sinful to be angry with Babamukuru.'[91] Throughout the novel, Tambu exercises these self-disciplinary measures upon her own person while watching the radical behaviour of her cousin accelerate.

The only occasion that Tambu nears transgression is at her mother and father's wedding, and the form of her subversion imitates Nyasha's. When Babamukuru condemns Tambu's parents' un-Christian union, he decides that they should be married immediately. Tambu feels invalidated by the nullification of her parents' relationship and refuses to attend the ceremony. In addition to the moral indignation she feels, Tambu's body 'reacted in a very alarming way.'[92] She suffers from a headache and an upset stomach; her bowels threaten to relieve themselves. The incident leads to an out-of-body experience: defying Babamukuru's orders causes Tambu to split in two: her mind floats out of her body. She calls this her 'newly acquired identity.'[93]

> He [Babamukuru] did not know how my mind had raced and spun and ended up splitting into two disconnected entities that had long, frightening arguments with each other, very vocally, there in my head, about what

ought to be done, the one half manically insisting on going, the other half equally manically refusing to consider it.[94]

Tambu's internal conflict signals a shift toward Nyasha's politics. This turn, however, does not last long. When she is offered the opportunity to enrol at Sacred Heart School in the next chapter, she banishes these thoughts. Tambu reverts to pursuing the narratives of progress as represented by the mission's Christian school. She is blinded by the promise of an English education. Without reservation, she throws herself into winning the spot at school. However, it is not long before this narrative is interrupted by the drama of Nyasha's eating disorder.

It is at this point that the overarching coherency of the narrative is disrupted, and it takes a disjointed turn. Once Tambu's burgeoning critical awareness is waylaid by Nyasha, her vocabulary takes on the same naïve quality that it had at the beginning of the novel: 'Excitement. Anticipation. Elation and exultation.'[95] The Sacred Heart School – run by nuns and attended by white children – represents colonialism's civilizing discourse, which Tambu accepts unquestioningly. After winning a place, she tours the school and sees the superior school library:

> Most importantly, most wonderfully, there was the library, big, bright, walled in glass on one side and furnished with private little cubicles where you could do your homework [...] I resolved to read every single one of those informative volumes from the first page to the last. With all those new books, reading took up so much of my time that there was none left in which to miss Nyasha, or my uncle and aunt.[96]

As Tambu is filling herself up with words, Nyasha does the same (while simultaneously emptying herself of food) somewhere off-stage. The largest portion of Nyasha's physical and psychological deterioration occurs while Tambu's narration drives toward the ideals of colonial development and education – the very ideas that the novel critiques through Nyasha's hunger. Nyasha is curiously absent from the narrative during this time. Tambu hears of her cousin's increasingly dangerous dietary regimen through letters from home but fails to consider it in any depth in her narration, until the very final chapter when she returns to her uncle's house.

Tambu displaces the onus of transgression upon the starving body of her cousin. It is Nyasha's narration that spurs on Tambu's transformation; it is Nyasha's tragic fate that causes 'seeds to grow.' Tambu is in the distanced position of voyeur, and. as such, she gleans the life lessons taught by Nyasha's hunger, but none of the bodily harm. Tambu's narration of the novel can be read as the bulimia she is too afraid to perform; instead of purging food, she purges words. She contains her cousin in the teleology of her realist narrative and learns a valuable lesson by watching her starve. This tempers her idealistic faith in the educational system of the colonial state. She uses Nyasha's violent protest to move from naivety to maturation – but with it come caveats. 'She develops into the modernized voice capable of recording the stories of the women around her but that is also curiously unable to represent her own developed body.'[97] Although Tambu productively reads her cousin's narrative and body in order to form her own authentic voice – as a postcolonial subject and a woman – there is a lack of lived experience in her own story. There is a kind of narrative vampirism in the text, whereby Tambu's narrative containment and digestion of Nyasha's body and its meanings allows for an uncomplicated negotiation of complex Shona female identity. Nyasha's body is eaten up by Tambu's narration. Contained within the realist text, Nyasha's hunger is translated into a finite teleology, easily consumed and assimilated.

Tambu cannibalizes her cousin's experience to advocate the lesson that words, writing, and education are how 'seeds do grow' – as is proved by the production of the novel itself. Despite the apparent ambivalence that she communicates through the novel's various other characters, she is relatively unburdened by doubt:

> Although I was not aware of it then, no longer could I accept Sacred Heart and what it represented as a sunrise on my horizon. Quietly, unobtrusively and extremely fitfully, something in my mind began to assert itself, to question things and refuse to be brainwashed, bringing me to this time when I can set down this story.[98]

These confident words conclude the novel, and the tone provides a sense of absolute closure – and a lesson learned. 'I triumphed. I was not seduced,'[99] she claims.

Nyasha's protest can be read in several ambivalent ways. It motions toward discursive agency, it embodies an untold history of national hunger and colonial exploitation, and it facilitates a deconstructive reading of the Cartesian duality and its attendant binaries. However, the price of Nyasha's hunger is physical and mental suffering, further dislocation of the hybrid subject, and the pernicious forms of femininity taken to their absolute limits. Like Nimi and Michael K, Nyasha's somatic rebellion is fraught with complexities. However, we can see here how the starving body might be productively read by those who witness it and may attain the lessons of anti-colonial politics through its sacrifices. Nyasha attempts to starve her own body and relies too heavily upon words and language (those of her colonial masters) to resolve her anxieties, but Tambu mimics this process in her writing of Nyasha's body with her own words. In terms of politicizing the experiences of Nyasha and Tambu, it can be argued that the book ends up tripping over its own critical analysis of western education as a civilizing and benevolent force. Tambu sounds confident that she has internalized the lessons embodied by her cousin's starvation, but we can trouble this neat ending by reading it through the ambivalences expressed by the other female bodies in the novel. Reading the novel with the scepticism that Tambu acquires over the course of the narrative may ultimately be the lesson the reader is meant to glean from the novel. By questioning Tambu's escape from the homestead and the mission, the ending of the book opens into a space of possibility where criticality and caution can exist alongside the hard-won female agencies contained in the text and body.

Notes

1 UDI stands for Unilateral Declaration of Independence, where a government self-proclaims independence from its mother country in order to form an autonomous nation state, as Rhodesia did from Britain in 1964. Rhodesia was the colonial name for present-day Zimbabwe. I use the term 'Rhodesia' when referring to the country within the context of Dangarembga's text; in all other instances, I use the term 'Zimbabwe.'
2 A note on the usage of the term 'anorexia': this chapter examines a character in *Nervous Conditions* who engages in anorexic behaviours. I argue that, as a teenage Shona girl living in Rhodesia, Nyasha's food refusal is a protest. Anorexia has been interpreted as a form of protest as well as pathology. See Susan Bordo, 'Anorexia Nervosa: Psychopathology as the Crystallization of Culture', in *Women, Knowledge, and Reality: Explorations in Feminist Philosophy*, ed. by Anne Garry and Marilyn Pearsall (New York: Routledge, 1996); Leslie Heywood, *Dedication to Hunger: The Anorexic Aesthetic in Modern Culture* (California: University of California Press, 1996); and Andreé Dignon, *All of Me: A Fuller Picture of Anorexia* (Bern: Peter Lang AG International Academic Publishers, 2007). I use the term 'anorexia' as it pertains to the general understanding of the term in cultural studies that appropriates medical discourses in their apprehensions, although it is worth nothing that not all medicalized comprehensions of anorexia intersect with the findings outlined in cultural studies. Therefore, when referring to Nyasha's food habits in this monograph, I use the terms 'anorexia,' 'hunger strike,' and 'protest' interchangeably.
3 Shelia MacLeod, *The Art of Starvation: A Story of Anorexia and Survival* (New York: Schocke, 1982), p. 54.
4 Michael Watts, 'Entitlements or Empowerment? Famine and Starvation in Africa', *Review of African Political Economy*, 51 (1991), 9–26 (p. 10).
5 Susan George, *How the Other Half Dies* (Harmondsworth: Penguin, 1976), p. 16.
6 Independent Commission on International Humanitarian Issues, *Famine: A Man-Made Disaster?* (New York: Vintage Books, 1985), p. 24.
7 See Thomas Robert Malthus, *An Essay on the Principle of Population*, 7th edition, reprinted (London: Dent, 1973).
8 Amartya Sen, *Poverty and Famines: An Essay on Entitlements and Deprivation* (Oxford: Oxford University Press, 1983), p. 4.
9 See John Iliffe, *Famine in Zimbabwe* (Gweru: Mambo Press, 1990).
10 See Alex De Waal, *Famine that Kills: Darfur, Sudan* (New York: Oxford University Press, 2005), pp.3–19.
11 Iliffe, p. 5.
12 The novel makes the divide between the affluent white colonialists and the poorer Shona natives very clear. Although Babamukuru is also Shona, his climb out of poverty is facilitated by the association with the colonialists and white Rhodesians. The demarcating line between the haves and have-nots is generally drawn between the native and the colonialist in the novel, but a more accurate dividing line would be those who embrace the economic and cultural discourses of colonial ideology versus those who do not.
13 Tsitsi Dangarembga, *Nervous Conditions*, 2nd Edition (London: The Woman's Press, 1988, repr. Banbury: Ayebia Clarke, 2004), p. 6.
14 Deepika Bahri, 'Postcolonial Hungers', in *Food and Literature*, ed. by Gitanjali G. Shahani (2018) 335–352 (p. 337).

15 Dangarembga, p. 3
16 Michael Gelfand, *The Genuine Shona: Survival Values of an African Culture* (Salisbury: Mogambo Press, 1973), p. 9.
17 Dangarembga, p. 82.
18 Ibid., p. 49.
19 Ibid., p. 20.
20 Ibid., p. 21.
21 Ibid., p. 30.
22 Ibid., p. 29.
23 Ibid., p. 29.
24 Michael Watts, 'Heart of Darkness: Reflections on Famine and Starvation in Africa', in *The Political Economy of African Famine*, ed. by R.E Downs, Donna O. Kerner and Stephen P. Reyna (Philadelphia: Gordon, 1991), 23–71 (p. 25).
25 Clare Barker, 'Self-Starvation in the Context of Hunger: Health, Normalcy and the "Terror of the Possible" in Tsitsi Dangarembga's *Nervous Conditions*', *Journal of Postcolonial Writing*, 44.2 (2008), 115–125 (p. 115).
26 Dangarembga, p. 13.
27 Ibid., p. 33.
28 Barker, p. 116.
29 Dangarembga, p. 26.
30 Ibid., p. 16.
31 Deepika Bahri, 'Disembodying the Corpus: Postcolonial Pathology in Tsitsi Dangarembga's *Nervous Conditions*', *Postmodern Culture*, 5.1 (1994), 2–27 (p. 14).
32 Dangarembga, p. 18.
33 Barker, p. 122.
34 Maud Ellmann, *The Hunger Artists: Starving, Writing and Imprisonment* (London: Virago Press, 1993), p. 10.
35 Isabelle Meuret, *Writing Size Zero* (Brussels: P.I.E. Peter Lang, 2007), p. 50.
36 Brandon Nicholls, 'Indexing Her Digests', in *Emerging Perspectives on Tsitsi Dangarembga: Negotiating the Postcolonial*, ed. by Anne Elizabeth Willey and Jeanette Treiber (Trenton, NJ: Africa World Press, 2002), 99–137 (p. 104).
37 Sue Thomas, 'Killing the Hysteric in the Colonized House: Tsitsi Dangarembga's *Nervous Conditions*', *The Journal of Commonwealth Literature*, 27.1 (1992), 22–26 (p. 22). See also Supriya Nair, 'Melancholic Women: The Intellectual Hysteric(s) in *Nervous Conditions*', *Research in African Literatures*, 26.2 (1995), 130–139.
38 Dangarembga, p. 137.
39 Nicholls, p. 117.
40 Dangarembga, p. 1.
41 Bahri, p. 2.
42 Ibid., p. 10.
43 Heather Zwicker, 'The Nervous Collusions of Nation and Gender: Tsitsi Dangarembga's Challenge to Fanon', in *Emerging Perspectives on Tsitsi Dangarembga: Negotiating the Postcolonial*, ed. by Willey and Treiber (2002), 3–24 (pp. 14–15).
44 Susan Z. Andrade, 'Tradition, Modernity, and the Family: Reading the *Chimurenga* Struggle into and out of *Nervous Conditions*', in *Emerging Perspectives on Tsitsi Dangarembga: Negotiating the Postcolonial*, ed. by Willey and Treiber (2002), 25–60 (p. 55).
45 Megan Warin, *Abject Relations: Everyday Worlds of Anorexia* (New Brunswick, NJ: Rutgers University Press, 2010); René Girard and Mark R. Anspach, *Anorexia and Mimetic Desire* (East Lansing: Michigan State University Press, 2013); Richard A. O'Connor and Penny Van Esterik, *From Virtue to Vice: Negotiating Anorexia* (New York: Berghahn, 2015); Antonio Mancini, Silvia Daini, and Louis Caruana, *Anorexia Nervosa: A Multi-Disciplinary Approach: From Biology to Philosophy* (Hauppauge, N.Y: Nova Science Publishers, 2010); Julie Hepworth, *The Social Construction of Anorexia Nervosa* (London: Sage, 1999); MacSween, Morag, *Anorexic Bodies: A Feminist and*

Sociological Perspective on Anorexia Nervosa (London: Routledge, 1993); Rudolph M. Bell, *Holy Anorexia* (Chicago: University of Chicago Press, 1985); Joan Jacobs Brumberg, *Fasting Girls: The History of Anorexia Nervosa* (New York: Vintage Books, 2000); Helen Malson, *The Thin Woman: Feminism, Post-Structuralism and the Social Psychology of Anorexia Nervosa* (New York: Routledge, 1998); Paula Saukko, *The Anorexic Self: A Personal, Political Analysis of a Diagnostic Discourse* (Albany: SUNY Press, 2008), and Petra M. Bagley, Francesca Calamita and Kathryn Robson, *Starvation, Food Obsession and Identity: Eating Disorders in Contemporary Women's Writing* (Oxford: P.I.E. Peter Lang, 2018).
46 Susan Bordo, *Unbearable Weight: Feminism, Western Culture, and the Body* (Berkeley: University of California Press, 1993), p. 5.
47 Ibid., p. 8.
48 Susie Orbach, *Hunger Strike: The Anorectic's Struggle as a Metaphor for Our Age* (London: Faber and Faber, 1986), p. xvii.
49 Kim Chernin, *The Hungry Self: Women, Eating and Identity* (New York: Harper Perennial, 1985), p. 47.
50 Nieves Pascual, 'Depathologizing Anorexia: The Risks of Life Narratives', *Style*, 35.2 (2001), 341–352 (p. 342).
51 Gilles Deleuze and Claire Parnet, *Dialogues* (New York: Columbia University Press, 1987), p. 110.
52 Sheila Lintott, 'Sublime Hunger: A Consideration of Eating Disorders Beyond Beauty', *Hypatia*, 18.4, (2003), 65–86 (p. 80).
53 Neela Ahuja, 'Colonialism', in *Gender: Matter*, ed. by Stacey Alaimo, Macmillan Interdisciplinary Handbooks (Farmington Hills, MI: Gale, 2017), 237–252 (p. 243)
54 Ellmann, p. 14.
55 Dangarembga, p. 59.
56 Ibid., p. 52.
57 Ibid., p. 64.
58 Ibid, p. 84.
59 Dangarembga repeatedly employs the words-as-food metaphor throughout the novel. See Dangarembga, p. 24: 'He thinks that because he has chewed more letters than I have, he can take over my children'; p. 36: 'having devoured English letters with a ferocious appetite! Did you think degrees were indigestible!'; p. 142: 'Because she is rich and comes here and flashes her money around, so you listen to her as though you want to eat the words that come out of her mouth.'
60 *The Holy Bible Containing the Old and New Testament, Authorized King James Version* (Victoria: Bible Protector, 2007), Deuteronomy 8:3.
61 Jonathan Brunberg-Kraus, '"Not by bread alone…": The Ritualization of Food and Table Talk in the Passover Seder and in the Last Supper', *Semeia*, 86 (1996), 165–191 (p. 165).
62 Jason Scott-Warren and Andrew Zurcher, 'Introduction' in *Text, Food and the Early Modern Reader: Eating Words*, ed. by Jason Scott-Warren and Andrew Zurcher (London: Routledge, 2019).
63 Deborah Lupton, *Food, the Body, and the Self* (London: Sage Publications, 1996), p. 18.
64 Ellmann, p. 22.
65 Dangarembga, p. 82.
66 Ibid., p. 107.
67 Ibid., p. 110.
68 Ibid., p. 193.
69 Ibid.
70 Ellmann, p. 17.
71 Dangarembga, p. 201.
72 Ibid., p. 202.

73 Ibid., p. 205.
74 Ibid.
75 Ibid.
76 Richard A. Gordon, *Anorexia and Bulimia: Anatomy of a Social Epidemic* (Oxford: Blackwell, 1990), p. 131.
77 Ellmann, p. 2.
78 Dangarembga, p. 205.
79 See Pauline Ada Uwakweh, 'Debunking Patriarchy: The Liberational Quality of Voicing in Tsitsi Dangarembga's *Nervous Conditions*,' *Research in African Literatures*, 26.1 (1995), 75–84; Sheena Patchay, 'Transgressing Boundaries: Marginality, Complicity and Subversion in *Nervous Conditions*', *English in Africa*, 30.1 (2003), 145–155; and Janice E. Hill, 'Purging a Plate Full of Colonial History: The *Nervous Conditions* of Silent Girls', *College Literature*, 22.1 (1995), 78–90.
80 Ellmann, p. 17.
81 Bordo, *Unbearable Weight*, p. 8.
82 Dangarembga, p. 1.
83 Ibid., p. 208.
84 Ibid., p. 58.
85 Ibid., p. 58.
86 Ibid., p. 57.
87 Ibid., p. 59.
88 Ibid., p. 30.
89 Ibid., p. 154.
90 Ibid., p. 131.
91 Ibid., p. 151.
92 Ibid., p. 151.
93 Ibid., p. 171.
94 Ibid., p. 169.
95 Ibid., p. 195.
96 Ibid., p. 200.
97 Ann Elizabeth Willey, 'Modernity, Alienation and Development: *Nervous Conditions* and the Female Paradigm,' in *Emerging Perspectives on Tsitsi Dangarembga: Negotiating the Postcolonial*, ed. by Willey and Treiber (2002), 61–82 (p. 70).
98 Dangarembga, p. 208.
99 Ibid., p. 70.

Bibliography

Ahuja, Neela, 'Colonialism', in *Gender: Matter*, ed. by Stacey Alaimo, Macmillan Interdisciplinary Handbooks (Farmington Hills, MI: Gale, 2017), 237–252

Andrade, Susan Z., 'Tradition, Modernity, and the Family: Reading the Chimurenga Struggle into and out of Nervous Conditions', in *Emerging Perspectives on Tsitsi Dangarembga: Negotiating the Postcolonial*, ed. by Ann Elizabeth Willey and Jeanette Treiber (Trenton, NJ: Africa World Press, 2002), 25–60

Bagley, Petra M., Calamita, Francesca and Robson, Kathryn, *Starvation, Food Obsession and Identity: Eating Disorders in Contemporary Women's Writing* (Oxford: P.I.E. Peter Lang, 2018)

Bahri, Deepika, 'Disembodying the Corpus: Postcolonial Pathology in Tsitsi Dangarembga's *Nervous Conditions*', *Postmodern Culture*, 5:1 (1994), 2–27

Bahri, Deepika, 'Postcolonial Hungers', in *Food and Literature*, ed. by Gitanjali G. Shahani (Cambridge: Cambridge University Press, 2018), 335–352

Barker, Clare, 'Self-starvation in the Context of Hunger: Health, Normalcy and the "Terror of the Possible" in Tsitsi Dangarembga's *Nervous Conditions*', *Journal of Postcolonial Writing*, 44:2 (2008), 115–125

Bell, Rudolph M., *Holy Anorexia* (Chicago: University of Chicago Press, 1985)

Bordo, Susan, 'Anorexia Nervosa: Psychopathology as the Crystallization of Culture', in *Women, Knowledge, and Reality: Explorations in Feminist Philosophy*, ed. by Anne Garry and Marilyn Pearsall (New York: Routledge, 1996)

Bordo, Susan, *Unbearable Weight: Feminism, Western Culture, and the Body* (Berkeley: University of California Press, 1993)

Brumberg, Joan Jacobs, *Fasting Girls: The History of Anorexia Nervosa* (New York: Vintage Books, 2000)

Brunberg-Kraus, Jonathan, '"Not by bread alone…": The Ritualization of Food and Table Talk in the Passover Seder and in the Last Supper', *Semeia*, 86 (1996), 165–191

Chernin, Kim, *The Hungry Self: Women, Eating and Identity* (New York: Harper Perennial, 1985)

Dangarembga, Tsitsi, *Nervous Conditions*, 2nd Edition (London: The Woman's Press, 1988, repr. Banbury: Ayebia Clarke, 2004)

Deleuze, Gilles and Parnet, Claire, *Dialogues* (New York: Columbia University Press, 1987)

De Waal, Alex, *Famine that Kills: Darfur, Sudan* (New York: Oxford University Press, 2005)

Dignon, Andreé, *All of Me: A Fuller Picture of Anorexia* (Bern: Peter Lang AG International Academic Publishers, 2007)

Ellmann, Maud, *The Hunger Artists: Starving, Writing and Imprisonment* (London: Virago Press, 1993)

Gelfand, Michael, *The Genuine Shona: Survival Values of an African Culture* (Salisbury: Mogambo Press, 1973)

George, Susan, *How the Other Half Dies* (Harmondsworth: Penguin, 1976)

Girard, René and Anspach, Mark R., *Anorexia and Mimetic Desire* (East Lansing: Michigan State University Press, 2013)

Gordon, Richard A., *Anorexia and Bulimia: Anatomy of a Social Epidemic* (Oxford: Blackwell, 1990)

Hepworth, Julie, *The Social Construction of Anorexia Nervosa* (London: Sage, 1999)

Heywood, Leslie, *Dedication to Hunger: The Anorexic Aesthetic in Modern Culture* (California: University of California Press, 1996)

Hill, Janice E., 'Purging a Plate Full of Colonial History: The *Nervous Conditions* of Silent Girls', *College Literature*, 22:1 (1995), 78–90

Iliffe, John, *Famine in Zimbabwe* (Gweru: Mambo Press, 1990)

Independent Commission on International Humanitarian Issues, *Famine: A Man-Made Disaster?* (New York: Vintage Books, 1985)

Lintott, Sheila, 'Sublime Hunger: A Consideration of Eating Disorders Beyond Beauty', *Hypatia*, 18:4 (2003), 65–86

Lupton, Deborah, *Food, the Body, and the Self* (London: Sage Publications, 1996)

MacLeod, Shelia, *The Art of Starvation: A Story of Anorexia and Survival* (New York: Schocke, 1982)

MacSween, Morag, *Anorexic Bodies: A Feminist and Sociological Perspective on Anorexia Nervosa* (London: Routledge, 1993)

Malson, Helen, *The Thin Woman: Feminism, Post-Structuralism and the Social Psychology of Anorexia Nervosa* (New York: Routledge, 1998)

Malthus, Thomas Robert, *An Essay on the Principle of Population*, 7th Edition, reprinted (London: Dent, 1973)

Mancini, Antonio, Daini, Silvia and Caruana, Louis, *Anorexia Nervosa: A Multi-Disciplinary Approach. From Biology to Philosophy* (Hauppauge, NY: Nova Science Publishers, 2010)

Meuret, Isabelle, *Writing Size Zero* (Brussels: P.I.E. Peter Lang, 2007)

Nair, Supriya, 'Melancholic Women: The Intellectual Hysteric(s) in Nervous Conditions', *Research in African Literatures*, 26. 2 (1995), 130–139.

Nicholls, Brandon, 'Indexing Her Digests', in *Emerging Perspectives on Tsitsi Dangarembga: Negotiating the Postcolonial*, ed. by Anne Elizabeth Willey and Jeanette Treiber (Trenton, NJ: Africa World Press, 2002), 99–137

O'Connor, Richard A. and Van Esterik, Penny, *From Virtue to Vice: Negotiating Anorexia* (New York: Berghahn, 2015)

Orbach, Susie, *Hunger Strike: The Anorectic's Struggle as a Metaphor for Our Age* (London: Faber and Faber, 1986)

Pascual, Nieves, 'Depathologizing Anorexia: The Risks of Life Narratives', *Style*, 35. 2 (2001) 341–352

Patchay, Sheena, 'Transgressing Boundaries: Marginality, Complicity and Subversion in Nervous Conditions', *English in Africa*, 30. 1 (2003), 145–155

Saukko, Paula, *The Anorexic Self: A Personal, Political Analysis of a Diagnostic Discourse* (Albany: SUNY Press, 2008)

Scott-Warren, Jason and Zurcher, Andrew, 'Introduction' in *Text, Food and the Early Modern Reader: Eating Words*, ed. by Jason Scott-Warren and Andrew Zurcher (London: Routledge, 2019), 1–16.

Sen, Amartya, *Poverty and Famines: An Essay on Entitlements and Deprivation* (Oxford: Oxford University Press, 1983)

The Holy Bible Containing the Old and New Testament, Authorized King James Version (Victoria: Bible Protector, 2007)

Thomas, Sue, 'Killing the Hysteric in the Colonized House: Tsitsi Dangarembga's *Nervous Conditions*', *The Journal of Commonwealth Literature*, 27. 1 (1992), 22–26

Uwakweh, Pauline Ada, 'Debunking Patriarchy: The Liberational Quality of Voicing in Tsitsi Dangarembga's *Nervous Conditions*', *Research in African Literatures*, 26. 1 (1995), 75–84

Warin, Megan, *Abject Relations: Everyday Worlds of Anorexia* (New Brunswick, NJ: Rutgers University Press, 2010)

Watts, Michael, 'Entitlements or Empowerment? Famine and Starvation in Africa', *Review of African Political Economy*, 51 (1991), 9–26

Watts, Michael, 'Heart of Darkness: Reflections on Famine and Starvation in Africa', in *The Political Economy of African Famine*, ed. by R.E. Downs, Donna O. Kerner and Stephen P. Reyna (Philadelphia: Gordon, 1991), 23–71

Willey, Ann Elizabeth, 'Modernity, Alienation and Development: *Nervous Conditions* and the Female Paradigm', in *Emerging Perspectives on Tsitsi Dangarembga: Negotiating the Postcolonial*, ed. by Ann Elizabeth Willey and Jeanette Treiber (Trenton, NJ: Africa World Press, 2002), 61–82

Zwicker, Heather, 'The Nervous Collusions of Nation and Gender: Tsitsi Dangarembga's Challenge to Fanon', in *Emerging Perspectives on Tsitsi Dangarembga: Negotiating the Postcolonial*, ed. by Ann Elizabeth Willey and Jeanette Treiber (Trenton, NJ: Africa World Press, 2002), 3–24

5 Traumatic National Hungers and the Starving Irish Body

Bobby Sands' 1981 Hunger Strike

On 1 March 1981, Irish Republican Army (IRA) prisoner Bobby Sands began a hunger strike in Long Kesh prison, just outside Belfast, that culminated in his death 66 days later. The hunger strike was to reclaim 'special category' status for IRA prisoners incarcerated for what they perceived as a war of liberation from the British colonial forces in the region. Previously, the special category status had provided recognition of and leniency for political prisoners to distinguish them from ordinary prisoners. It had been revoked years earlier in a bid to delegitimize and clamp down on escalating sectarian violence over the period of conflict in Northern Ireland known as the Troubles.[1] Along with Sands, many other Republican prisoners signed up for the hunger strike. It caused a furore of publicity and for eight months the events in Northern Ireland would capture the world's attention. Media coverage documented the standoff between British Prime Minister Margaret Thatcher and the young IRA men who were desperate and daring enough to gamble with their lives rather than be branded as terrorists and criminals.

Sands was born into a Catholic family in 1954. When he was 18, his family moved from their home in Rathcoole to the Twinbrook housing estate in West Belfast due to Loyalist intimidation. In 1972, Sands joined the Provisional IRA, and in October of the same year he was charged with the possession of firearms and sentenced to prison for five years, but was released three years later. He immediately resumed his IRA activities upon release and was again imprisoned, for 14 years, for possession of firearms in 1977. While in prison, Sands wrote profusely. His writing is politically didactic. His poetry is structurally strict and follows tight rhyme schemes and beat structures, mirroring the sense of enclosure and monotony he experienced in Long Kesh. In contrast, his prose pieces are often chaotic and rambling. The juxtaposition of these forms reflects the state of his life in prison: tightly structured tedium and endless time. His writing subscribes to and reflects a long-standing tradition of anti-British sentiment as reflected in Republican and Nationalist anti-colonial discourses.

The hunger strike concluded with the death of ten prisoners; with most of the concessions ultimately attained, excepting the granting of political-prisoner status, it was a partial victory for both sides. The story of the hunger strike appears straightforward, with clear political dividing lines between the two

DOI: -5

parties involved. However, the mechanisms, motivations, and realities of the hunger strike were much more complex. The story of Ireland, and its historical, political, and social difficulties with its English governors, is a long and turbulent narrative. This chapter analyzes the Irish hunger strike as a response to a culmination of complicated historical legacies and the continuities of these contexts into contemporary Northern Irish politics. The complicated interweaving of these historical parameters both found expression in and affected Sands' writing, Nationalist ideology, and starving body. I contend that Sands' hunger strike was a contemporary response to ideological, historical, psychological, physical pressures that produced contradictions. These contradictions were inscribed on his body within the context of internment, and the hunger strike was the method he used to try to resolve them. I examine Sands' prison writing to explore the historical and contemporary contexts that structured his time in prison and his hunger strike. I also consider how these very narratives contributed to the ambivalences produced by his protest and evaluate its political successes and failures.

Sands' prison strike differs from the other hungers studied in this work in key ways. His hunger strike was an overtly political protest and anti-colonial strategy. Like Nimi, Michael K, and Nyasha, Sand's hunger is situated within a discursive logic of signification. Sands' hunger is a somatic form of speech. However, unlike the other hungry bodies in this study, Sands produced a wealth of words explaining, documenting, and rationalizing his strike. He is not the silent hunger striker of past chapters, filling himself up with words instead of food, or refusing both. He consumes language, words, and writing, and produces it – particularly about his internment, hunger strike, and eventually, his imminent death. Sands' hunger narration is a first-person account, whereas in previous chapters the hungry body has been recounted by a second- or third-person narrator. Sands' body speaks for itself, repeatedly and almost obsessively. The overtly political nature of the hunger strike within the Irish context produced a deluge of words, both on Sands' behalf and by Sands himself. For Sands, excessive writing was an effort to rewrite the meanings of the body – not merely to starve meaning out of the body, but to wilfully reassert a different narrative through its hungers. Alternative forms of silence do figure into my analysis of Sands, but rather than emanating from the individual's starving body, the silence I consider is the historical exegesis around the Great Irish Famine, which serves as the historical matrix for my reading of Sands' protest. The trauma associated with the historical recording (or lack thereof) of the famine, as well as the material and psychological impact it left on Ireland, will frame my reading of Sands' hunger strike. Despite the differences between Sands and the other bodies in this work, I examine the subtle ways that Sands was also unable to represent a stable sense of self – despite the prolificacy of his prison writing – produced by contradictions contained within his hunger.

I consider historiographical narratives that set the scene for an enduring Irish–British enmity that produced the nationalist ideology used by Sands to articulate his strike. I examine a selection of Sands' prison writings for the interplay of these historical processes with his own contemporary contexts.

Comparable to hunger strikers in previous chapters, Sands articulated his hunger strike through a mind–body dualism. Thus, his bodily resistance is also read as a desire to resolve the complex and contradictory subject position manifested in the colonial/postcolonial body. Sands' writing illuminates the paradoxes contained in his protest and situates his hunger strike within this larger historiography. Both events construct the starving Irish body within the context of a historical metanarrative of colonial relations between England and Ireland. This comparative reading considers the traumatic repetition of forms of hunger and starvation, and examines how these hungers are inserted within a historiographic tradition that itself is polemicized by the politics of representation and testimony.

A Great Hunger: Trauma, Memory, and Irish Nationalism

Historians and sociologists agree that conditions preceding the Great Famine indicated that Ireland was vulnerable to a subsistence crisis. Widespread poverty coupled with a sharp spike in population from the middle of the eighteenth century onward set the stage for a substantial food shortfall. Ireland was plagued by extreme poverty that was contextualized by centuries of imperial exploitation and occupation; in the immediate historical context of the famine, this mainly played out in the arena of land ownership. More than a quarter of the Irish population were landless labourers, either bound and contracted to work for wealthier landowners in exchange for housing and a percentage of profits or forced to roam around the country looking for work and subsistence.[2]

The remainder of Ireland's land resources were disproportionately divided between a smallholding demographic, known as cottiers, and wealthy landowners – predominantly members of a 'Protestant ascendancy' of Anglo-Irish or British origin who mostly resided abroad as absentee landlords and managed their affairs through middlemen. With the added pressure of the increase in population, the cottier class suffered further subdivisions:

> Under the acute pressure of the population explosion the subdivisions of holdings had been carried to extraordinary lengths by the eve of the famine. Aside altogether from the 135,000 holdings of less than 1 acre in 1844, almost half of the other 770,000 holdings did not exceed 10 acres, while another quarter were between 10 and 20 acres.[3]

As a result, productivity was low, population growth drove up the prices of land and rents, and this worsened unemployment. These factors meant that food subsistence had already been an ongoing problem in Ireland, with poor crop yields in past years having caused substantial numbers of deaths.[4] These shortfalls have since been relegated to a more minor role in popular historical narratives in the light of the Great Famine of 1845. What emerges from this picture is that most of the Irish population were extremely poor and living in ongoing food-insecure conditions on the eve of the famine.

Exacerbating this threat even further was the reliance on the ubiquitous potato by most landless and low-income families. High in yield, convenient, and fairly nutritious, the potato was an affordable solution to the nutritional needs of over half of the 8.5 million Irish population.[5] Despite the lack of variety in the Irish diet, the potato proved itself a hearty and nourishing crop, and on it the Irish labourer thrived: 'Research into the height of Irishmen born before and after 1815 indicates that adult Irish males were taller than their English peers and by implication were reared on a healthier diet.'[6] Ireland's reliance on the crop contributed to population growth. It was cheap and easy to cultivate, allowing people to marry earlier and support a growing family.[7] So although the Irish were seemingly indentured to the potato, it provided, on the whole, a good source of nutrition that allowed the Irish population to flourish in both health and numbers. It is this very exclusivity, however, that meant the arrival in 1845 of a particularly damp harvest season, and the fungus *Phytophthora infestans*, proved catastrophic. The massive failure of the crop in 1845, and then again in 1847, followed by a series of ineffectual relief programs, caused a devastating loss of life over the course of the famine. Over the seven-year ordeal, starvation, disease, and mass emigration desolated the Irish population. Although the question of statistical accuracy on the subject of the mortality rate has proven to be a contentious one, two independent studies – by Mokyr[8] and by Boyle and Ó Gráda[9] – suggest the figure is around one million excess deaths, or over one-ninth of the population. This figure, moreover, solely encompasses the effects on population due to starvation-related death; the Irish exodus to North America and Australia, combined with famine-related disease (which generally claims far more lives during famine conditions than starvation itself), roughly doubles that figure.

Many critics read the Great Famine and its place in Irish memory through a Trauma Studies lens.[10] Despite the exceptional fatalities caused by the famine, the initial critical exploration of it and its causes was relatively mute. The journal *Irish Historical Studies*, founded in 1938, published only five articles related to the famine in the first 50 years of its existence.[11] A Trauma Studies approach reads the Great Famine as an event so shattering to the national psyche that it has been repressed and unprocessed – thus explaining the initial historiographical silence. Traumatic events are problems of narrativization. The traumatic experience is so deeply wounding that it remains lodged in the present and cannot be enunciated in a temporally stable past tense. It escapes the bounds of representation – whether it be historical, linguistic, artistic, or even cognitive. 'Trauma is, by definition, not capturable through representation or, indeed, recollection; it is known through the gap that disrupts all efforts at narrative reconstruction.'[12] The Great Famine has been compared to other traumatic historical events in terms of scale and subsequent silences and repressions:

> The Irish Famine was first linked to the Holocaust by Frank O'Connor in a review of Cecil Woodham-Smith's *The Great Hunger* in 1962; whatever view critics may take of the aptness of such controversial comparisons, studies of Famine literature have been dominated by a 'parallel between

the problems of artistic representation presented by the Famine and those presented by the Nazi Holocaust.'[13]

Like the Holocaust, the losses produced by the Great Famine escape the bounds of discourse. It is uncontainable in language – as Margaret Kelleher asks: 'is it possible to depict the horror and scale of an event such as famine, are literature and language adequate to the task?'[14] The inadequacy of representation in this context refers to the deaths of witnesses whose memories are forever lost, and the failure of language to capture what actually was experienced and remembered by those who survive:

> Much of knowing is dependent on language [...] Because of the radical break between trauma and culture, victims often cannot find categories of thought or words for their experience. That is, since neither culture nor experience provide structures for formulating acts of massive aggression, survivors cannot articulate trauma, even to themselves.[15]

David Lloyd theorizes that the famine has become 'an excess of indigence that plagues the canons of narrative representation.'[16] In the field of Irish famine writing, this excess is mediated in the form of a stringently formulated historiography, didactic in nature and politically charged. The initial silence in scholarship about the Great Famine was followed by a later academic outpouring in the mid-1990s. Before this moment, the void of scholarship was filled by histories that were divided into two separate groups with differing ideological variations. The two main discursive camps in this debate are Nationalists and Revisionists. This subsequent history writing on the Great Famine documents an ongoing political controversy that still saturates critical theory explaining (a) *why* the famine occurred and (b) whether it could have been avoided, or some of its devastating effects ameliorated. Revisionists and Nationalists take opposing viewpoints on the matter.

Popularized by Cecil Woodham-Smith's 1962 study *The Great Hunger* – a hugely popular work that holds the honour of being the most widely read Irish history book of all time – Nationalist accounts of the Great Famine 'laid responsibility for mass death and mass emigration at the door of the British government, accusing it of what amounted to genocide.'[17] This is necropolitical policy–making realized. Nationalist accounts are critical of British relief policies, appraising them as grossly unfair and malicious. They present Britain's depiction of Irish labourers as essentially lazy and incompetent. This racist depiction of events shores up and attributes responsibility for the famine to Ireland and the Irish themselves. Nationalists point out that relief policies during the famine accomplished what had dogged the landowning class for so long – the clearing of smallholders and the consolidation of subdivided land – and argue that this was the intended political aim all along. Nationalist discourses give credence to figures such as John Mitchel, a Young Irelander whose damning work *The Last Conquest of Ireland (Perhaps)* lays blame squarely on

England's doorstep.[18] In equal measure, these discourses vilify Sir Charles Trevelyan, who was in charge of the British relief programme and is often condemned for his role in the poor management of the crisis. Today, the sentimentality associated with Woodham-Smith's *The Great Hunger* renders the work a somewhat controversial academic text. Considerations of bias and the socio-political context in which it was written, and the Nationalist tradition commonly associated with it, means that the text has fallen out of academic favour. It still has value, however, as Colm Tóibín reminds us: 'Nobody will be able to write like that again. Reading *The Great Hunger* is like reading Georgian poetry while knowing that a new, fractured, "modern" poetics is on the way.'[19] *The Great Hunger*'s reception and popularity can instruct us on the politics of famine writing, and reminds us that the Nationalist tradition it emerges from is far from being obsolete.

The professionalization of Irish history in the twentieth century brought with it an academic trend of 'revising' popularly accepted Irish histories. Revisionists claimed that Nationalist narratives sensationalized Irish history and perpetuated populist and reductive narratives. Regarding the Great Famine, they particularly aimed to debunk the image of Ireland as absolute victim of British imperialism. Instead, they claimed to present more nuanced and empirical modes of thinking and writing.[20] These historical narratives were also accused of harbouring specific political aims, despite the purported rationality of their history-making. 'Revisionist politicians and historians, anxious to escape a binary relation with Britain that placed them constantly in a position of inferiority, sought to develop a nationalist grand narrative for Ireland.'[21] Revisionism presented an alternative route to nation-building. It advocated for a nationalism that rewrote Irish victimhood in favour of moving forward into a heroic, idealized national future. These 'Revisionist' accounts, however, were accused of producing analysis in an overly empirical rhetoric that managed to 'filter out the trauma in the really catastrophic episodes of Irish history, such as the English conquest of the sixteenth century, the great rebellion of the 1640s, and the great famine itself.'[22] With regard to the famine, these revised historical accounts attempted to invalidate the familiar Nationalist accusation that the British had systematically and malevolently allowed or exacerbated vulnerable political and social conditions, thus causing the maximum fatalities during the famine. Instead, they championed what they believed was a contextually sensitive understanding of British policy – particularly in the assessment of Britain's decision to place the entire economic onus of relief on to Irish landlords in the form of the ill-fated revised Poor Law of 1847. Revisionist narratives expressed a sincere belief in the economic doctrine of laissez-faire as the best means for providing adequate relief. Nationalist narratives, claim Revisionists, 'failed to see the Famine for what it was: a mere catalyst of changes that were on the way in any case.'[23]

More recent scholarship[24] has acknowledged the need for a more diversified picture of events leading up to and during the famine, and focuses on the interplay between multiple factors of causation and context. For example, Nationalists, in their insistence on a deadly British intent, have often argued that Britain could have averted the famine by closing ports to exports. There was indeed a food shortfall owing to the potato famine, but studies have concluded that this gap would not have been alleviated by the closing down of exports during this period: 'The food gap created by the loss of the potato in the late 1840s was so enormous that it could not have been filled even if all the grain exported in those years had been retained in the country.'[25] This reading has been glossed over by strictly Nationalist approaches that employ a more entitlement-based methodology to accuse the British of wilful negligence. Further still, recent scholarship gives credit to British policy makers who seemingly had genuine intentions with regard to relief measures. British Prime Minister Robert Peel's sincere efforts to deter the worst effects of the crop failure of 1845 have been noted by critics, and his inevitable failure to get the requisite bills put through parliament is viewed less as malicious intent, but more of a failure of a political system that was disproportionately controlled by interested landed parties. This type of critical work tries to steer clear of dichotomizing Nationalist and Revisionist accounts and situates itself instead in the 'grey zone'[26] – a term used by Primo Levi to describe the Holocaust – to complicate the stringent binary between victim and perpetrator of violence.[27] Often using testimonials and individuated accounts, this perspective atomizes the famine experience to its most basic components, exposing the brutality of lived experience. As Breandán Mac Suibhne pleads: 'Genocide-criers, genocide-deniers, do us all a favour: just tell us what the Famine was like for the cottier and the labourer, the small farmer and the artisan.'[28] These forms of narrative also:

> acknowledge some issues raised by the condition to which ragged humanity was, in places, reduced in Ireland in the late 1840s, and [...] bring into view situations in which moral judgment may not then have been impossible but can scarcely have been easy, and today seems utterly inappropriate.[29]

These histories instead contextualize the difficult parts of the famine and interrogate didactic narratives that paint the Irish as parochial victims and all landlords as malevolent villains.[30] These accounts do not, by any means, excuse the British or qualify the famine as a natural progression of providential factors (as many Revisionist histories do), discount the role of imperialism in Ireland as a breeding ground for socio-economic weaknesses that certainly exacerbated the effects of food shortfall, or dismiss British relief measures that could be read as Malthusian approaches to population control – inflected by cultural racism. They do, however, add detail to a landscape that between the 1950s and 1990s has historically been, and sometimes continues to be, characterized by an oversimplified split between Nationalist and Revisionist narratives.

Lloyd states: 'there is no singular memory of the past at all, whether historical or popular, let alone one of the famine.'[31] Although this is certainly the case, the politicized terrain of the Great Famine speaks to larger difficulties around representation. The various permutations and debates within its historiography speak to the problematic of capturing the Great Famine, a traumatic event of such intensity that memory and testimony become fragmented and diverge in different directions. Among the range of responses to trauma, narrativizing the event is considered the most useful way of processing. If trauma 'creates a speechless fright that divides or destroys identity,'[32] then a possible way through trauma is to 'recreate or abreact through narrative recall of experience.'[33] This resolution of the traumatic experience presents its own problems, however, as '[a]ccurate representation of trauma can never be achieved without recreating the event since, by its very definition, trauma lies beyond the bounds of "normal" conception.'[34] However, recreations of trauma are in turn unreachable because of the limit-shattering nature of the event, both at the time of its occurrence and in memory acts that follow. 'The ability to recover the past is thus closely and paradoxically tied up, in trauma, with the inability to have access to it.'[35] Therefore, although testimony and remembrance are read as ways of 'working through' trauma, they contain within them the very seeds of their own undoing. Furthermore, traumatic memory is often inaccessible to voluntary recall – memories shift and alter, and are subject to many distortions over time and in retellings.

How can history writing, with its emphasis on accuracy and materialism, approach the representational limits of the traumatic event? I propose that the traumatic act of fragmented remembrance plays out in the field of historical famine writing, with its competing depictions and metanarratives. '[I]f trauma is a crisis in representation, then this generates narrative *possibility* just as much as *impossibility*, a compulsive outpouring of attempts to formulate narrative knowledge.'[36] In the context of immense national trauma, historically containing the Great Famine becomes a politicized act – one structured by the stringent binaries of Nationalism and Revisionism, a 'moral black hole to trauma sucking in nuance and leaving only extremism beyond its pull.'[37] Although recent scholarly work on the Great Famine is more considered in its approach, Nationalist narratives of famine historiography still dominate the popular, public memory in Ireland.[38] They provide a clear representational order in response to the traumatic event. In literary terms, these kinds of historical accounts are akin to the realist novel in that they assume 'that the world itself is story shaped – that there is a narrative, sometimes of a consolingly teleological kind, implicit in reality itself.'[39] They shut down, rather than complicate, the occlusions and anxieties produced by the Great Famine. Lloyd states:

> Encrypted in nationalist culture, then, would be the loss not so much of a primordial and even pristine Gaelic culture as the loss of forms of agency and of social relations that cannot even fully be named – the loss, we might say, of the loss itself. This encryption, the burial without wake or memorial, of the loss of the loss of the dead generates the melancholy turn that is held to afflict post-Famine Irish culture.[40]

The vacuum of pain and loss left behind by the Irish famine is filled with simple dichotomizing narratives because the salve for the non-narratable trauma is the stringently narratable. Moreover, Nationalist narratives that are predicated on a romantic history of Gaelic culture, Irish language, mythic landscapes, and anti-colonial struggle – or the 'national trilogy of land, people and history'[41] – facilitate a simplified and idealized version of Ireland that is rooted in a politicized nativism. These histories respond to the 'representational challenge [...] heightened by a need to bridge the gap constituted by a relative silence on the Famine in earlier Irish literature, so as to come to terms with what is perceived as a hitherto repressed collective trauma.'[42] The project of Irish nation-building is focalized through a unified cultural and national identity: 'Irish nationalism, like its counterparts in other colonial nations, is a modernizing movement that seeks to deploy a selected and canonized version of the past in the service of the political and social projects of the modern state.'[43]

Irish Nationalism and Republicanism tapped into these romanticized visions of Ireland, which in turn fuelled anti-colonial resistance and writing. The 'loss of the loss itself' outlined by Lloyd is the unreachable trauma of the famine. Rather than trying to locate a 'real' version of events in the form of history writing – rendered impossible due to the nature of the trauma itself – it may be more productive to think of Nationalist narratives as a 'canon of existent medial constructions.'[44] The Great Famine is a powerful site of traumatic memory that has been repeatedly represented in a variety of media – in literature, films, artwork, histories, and journalism. These are representations that, although fraught with the pain, loss, and violence of the famine, are mediated receptions that can be read as a digestible discursive version of historical events – rather than the linguistically obliterating 'truth' of the famine itself. Cemented in the agon between Britain and Ireland, Irish Nationalism provided a clear path from past trauma to future wholeness, articulated in anti-colonial struggle. Nationalist representations of Ireland:

> have been premediated, that is, the already existing representations of these historical events have set the terms for their future representations. The Famine of 1845–48 is a good example from Irish history of premediated, of the dynamics of collective memory and forgetting [...] and of the perpetuation of certain versions of history.[45]

This is what Chris Morash calls a 'semiotic system of representations which had replaced the Famine.'[46] Irish Nationalism has been instrumental in Ireland's anti-colonial struggles, and perhaps where and how it is produced – the 'authenticity' of its mimetic capacity – is less important than its power as an ideological framework to structure political resistance against oppression:

> For the writers of The Nation [...] writing was much more reader-oriented, more dedicated to efficacy than accuracy, more an instrument for communicative action than a looking glass [...] The point was not necessarily to

'contain' the world of the Famine within language – it was primarily to write about the Famine in a way which would produce certain responses.[47]

These responses were underpinned by the Irish Nationalisms that emerged out of several traumas experienced in Ireland, including the Great Famine, and animated the anti-colonial struggles against the British colonial power.

It is worth bearing in mind the traumatic response of Nationalist discourses as a necessary defence against the trauma of hunger when considering the starving body of Bobby Sands. Even if Irish Nationalism (and in other cases Revisionist Nationalisms) had productively filled the void left by the larger historical traumas of colonialism that Ireland experienced, the essential traumatic dynamic of the famine can be read as still unprocessed. This understanding of Irish history provides a useful theoretical space to link the trauma of Sands' starving body to the Irish famine itself. Lloyd argues that an unspeakable, limit-shattering trauma lies at the heart of a national silence around, and repression of the memory of, the famine.[48] 'It is a spectre that continues to haunt because we are not yet free, collectively, of the shadow of processes of dehumanisation that have their counterparts all too evidentially in our own time.'[49] Lloyd's work posits a collective national psyche with a cohesive, and shared, memory. Michael O'Loughlin makes a similar case for the fragmented Irish psyche. Rather than reading through the history books for traces of the famine, he speaks of the individual and intergenerational silences around the Great Famine. He claims that unwillingness to revisit and transmit the traumatic events of the famine 'precluded whole generations of Irish children from engaging with this catastrophic loss that is embodied in every person of Irish descent.'[50]

> The core of trauma arising not from the traumatic memory per se, but from the disavowal of that memory, leaving the experience unsymbolized, hence inaccessible to the psychotherapeutic process Freud referred to as working through. Some analysts refer to such trauma as unspeakable or unsayable.[51]

In this understanding of the Great Famine, if the famine remains unprocessed and silenced, it will find a way of emerging uncontrollably from the cultural, social, and somatic body/bodies of Ireland. '[T]he Famine is inscribed in the national – and indeed – international memory of the Irish and the British in indelible ways.'[52] Sands' hunger strike can be read as an expression of the unspeakable hungers suffered by the Irish body during the Great Famine, and perhaps can be posited as a sort of 'working through' in Freudian terms, if not also a repetition of the traumatic event in an attempt to work through it. '[T]he impact of trauma can only adequately be represented by mimicking its forms and symptoms, so that temporality and chronology collapse, and narratives are characterized by repetition and indirection.'[53] This may explain the repeated return to performances of hunger in Ireland's history, particularly in the form of political hunger strikes. 'In traumatic foreclosure, anything unsymbolized is excluded, engineering its reappearance through symptoms.'[54] Sands' hunger can be read as a re-enactment of a repressed memory of the famine by using his own body as a site of violent repetition.

But this is not the only way to link the Irish famine and Sands' hunger strike.[55] Sands was also responding to his own contemporary milieu, which was largely produced by nationalisms forged from the trauma of famine, permutations of which animated Sands' own cultural present. 'Collective memory, informed by a simplistic understanding of the past, always tells us more about the present than the past.'[56] This chapter considers the ways in which the idea of the past – as it is produced in the present moment – is engaged with and negotiated in Sands' writing. Therefore, a consideration of the historiography of Irish famine writing is relevant – not as a narrative that effaces the 'real' trauma of the famine (and so should be discounted as an oversimplified concealing narrative that prevents psychological processing) but as a version of history that has been used to process and coherently narrativize the collective traumas of the nation. The history of the famine is being engaged with, but through a twentieth-century historical lens that no doubt owes much of its form and context to imperialism and the nationalisms it produced. Both the politics of the past and the present shaped the staging of Sands' protest. He responded to a tradition of narrativization of famine trauma that is nationalist and populist in nature – a firebrand, militant nationalism which persistently reappears in Nationalist scholarship around the famine as well.

The starving body is a contested space in several different, often-conflicting, causative historical narratives about the famine. 'If [h]istory, then, was a contested field where colonizers and colonized struggled for the right to give meaning,'[57] the starving body became the locus of these contestations. The starving body can be read as the target and result of wilful British malevolence, as the victim of a cruel political economy, or as the inevitable focus of a cruel providence.[58] Remarking upon the 150th Anniversary of the potato famine, Christine Kinealy notes:

> [T]he anniversary of the appearance of the potato blight and subsequent Famine has attracted a lot of public and media attention [...] The coincidence of the Northern Ireland peace process has increased international interest in Irish affairs, and a number of articles, TV and radio reports have seen the two events as being inextricably linked.[59]

Historical contestations were being rehearsed in and on Sands' starving body during the 1981 hunger strike. Bobby Sands consciously re-invoked the binaries that characterized, and which continue to characterize, the popular Irish Nationalisms that solidified after the Great Famine. According to these narratives, at the time of the famine, the colonial relationship between Britain and Ireland manifested itself in the starving bodies of the Irish peasantry – and it is this particular exegesis of the famine, which remains a dominant image in the popular Irish imaginary, that Sands most consciously engaged with. Sands used his own body to purposefully intervene in these discourses of imperialism, nationalism, and Anglo-Irish relations to restage an enmity that has characterized Irish national identity for decades. His hunger strike threw the body into crisis, the resolution of which might determine the starving body's – and thus Ireland's – occluded traumatic 'truth.' The degree to which this protest succeeded is examined in the latter sections of this chapter. But first I consider how Sands draws a conscious link between his own starvation and internment and a larger historical narrative of Anglo-Irish antagonism in his prison writing, re-articulating the stringent binaries that constituted both popular representations of the Troubles and Irish Nationalism.

Historical Hunger: Persistent Nationalisms and Bobby Sands' Prison Writing

In his book *Heathcliff and the Great Hunger: Studies in Irish Culture*, Terry Eagleton notes that the failure to indoctrinate the Irish population with British ideology is at the root of the combative relationship between the two cultures. The lack of British hegemonic inculturation in Ireland created problems in terms of governance:

> It is doubtful whether any governing bloc in history has been able to achieve such unqualified sovereignty over its subjects, as [Raymond] Williams is the first to insist; ruling classes have been on the whole more tolerated than admired. But there were particular problems in this respect in colonial Ireland. For the truth is that no occupying power in the country was able to attain a hegemony sufficiently widespread, enduring and well-founded for its ends.[60]

Ruling nations not only impose the material realities of an economy and government upon their subjects; rather, these structures install a pervasive ideological dominance within the very social structures of culture itself, to the extent that this dominant ideology becomes accepted and internalized within the minds of the subjugated class. The effect of this failure to instil hegemonic structures into the social and psychological body politic in Ireland, Eagleton argues, generated national ramifications that persist in the present.

Political expressions of Irish Nationalism and religiosity structured along the strict dichotomization of English and Irish (or Protestant and Catholic) intensified after the Great Famine.[61] As a non-settler colony, Ireland was administered by a small but militarily effective colonial presence, and this may go some way toward explaining a failure of colonial indoctrination, and Ireland's use of 'Englishness' to define what it is not. Irishness is forged in the agon between England and Ireland. 'So completely is the history of the one country the reverse of the history of the other that the very names which to an Englishman mean glory, victory and prosperity to an Irishman spell degradation, misery and ruin.'[62] Woodham-Smith's assertions are replicated in a variety of critical sources. 'Ireland was soon patented as not-England, a place whose peoples were, in many important ways, the very antithesis of their new rulers from overseas,'[63] says Declan Kiberd.

> It was the English insistence that Ireland was a barbarous country which drove the Irish to construct, so as to sustain their own pride, a romantic and consoling counter-image, in which ancient Ireland was displayed as a land of saints and scholars, warlike but chivalrous Celtic heroes.[64]

This dichotomized version of history produces simplistic readings of the past: 'postcolonial identities that are shaped around trauma may have a legacy of manichean [sic] vision that denies internal divisions in the name of a unity that was imposed by the colonial power.'[65] Irish anti-colonial thought, then, has tended to follow Benita Parry's theory:

> If it is conceded that the structure of colonial power was ordered on difference as a legitimating strategy in the exercise of domination, then it could be argued that the construct of binary oppositions retains its power as a political category.[66]

In this structuring of difference, the English are rejected, but continually – and necessarily – haunt the psychosocial limits of Irish subjectivity and history, simultaneously giving shape to the latter and troubling/reinforcing its boundaries by interfering with its sense of national independence and agency.

The dichotomizing nationalisms that structured anti-colonial struggle occluded the more complex histories from which they emerged:

> While [Irish Nationalism] seeks to forge from the past the continuity of a tradition in which the distinctiveness of Irish culture and therefore the legitimacy of its aspiration to separate nationhood are affirmed, nationalism is no less obliged to suppress and occlude those social and cultural elements among the people that are as recalcitrant to its version of modernity as they are to those of the colonial state.[67]

While Irish Nationalism was thus integrally structured on British antagonism, it also produced a sanitized version of Irish history, which was embedded in the public consciousness:

> Public memory is typically understood as animated by affect. That is, rather than representing a fully developed chronicle of the social group's past, public memory embraces events, people, objects and places that it deems worthy of preservation, based on some kind of emotional attachment.[68]

Many forms of cultural Irish Nationalism also reproduced the hegemonic culture of the English. Political aims presented as a replica of the English constitution based on the tenets of bourgeois property rights, patriarchy, and the heterosexual family as the universal ideal.[69] To deviate from this model was to fall out of the Enlightenment order that structured European character:

Continuously subjected to this Anglocentric imperial narrative of history as progress, the Irish nationalist intelligentsia of the 1840s found themselves struggling within a web of contradictions. Were they to claim [...] that Ireland was participating in that progress of which England boasted so loudly, they would have been denying the very thing which differentiated them from their colonizers, and which proved the failure of the colonial administration: to pronounce that 'the history of Ireland is the history of progress' would have been to admit the success of the Union. However, to distance Ireland from the entire complex of concepts that went into the mid-nineteenth-century formulation of the idea of progress would have been to deny that Ireland had a right to the political liberty which was the index of social progress.[70]

These contradictions presented problems for Irish Nationalism and their anti-colonial discourse and strategies.

Moreover, to forge a unified nationalist platform from which to deploy anti-colonial action, Irish Nationalism has at various points in history manufactured ways of eliding sectarian, religious, and cultural affiliations within a diverse population. Debunking the national mythology of the Celtic native, Richard English states: 'There was no single, original Gaelic or Irish race, just as there were no discernible natives in the sense of an original people than whom all others and their descendants are less truly Irish.'[71] These kinds of strategies have been deployed from several influential sources. Mitchel formulated his earlier theories of national belonging by tying citizenship with land: 'time conferred nationality. For settlers, familiarity bred citizenship. This enabled him to discount historical arguments about the source of property titles and offer a purely synchronic argument about exploitation.'[72] By overcoming the class and religious differences of his time, Mitchel found a means of consolidating anti-colonial struggle against a singular enemy – the English.[73] Within the context of the Troubles in Northern Ireland, political subdivisions and competing allegiances became even more pronounced.

> It was a British state which discriminated against you as part of a nationalist Irish community. And when the struggles became clarified in local battle lines (police from beyond your community clashing with neighbours from within it,) and when it was Nationalists who offered both explanation and seemingly powerful defence and remedy, it was nationalism which became the vehicle for your urgent struggle.[74]

After all, in Northern Ireland not all Nationalists were Republicans, seeking an independent, unified Irish state, although Republican politics does emphasize a community-based nationalism predicated on place, which necessarily required concessions between Unionists and Loyalists. But Irish Republicanism and its emphasis on nationalisms rooted in a common cultural, historical, and linguistic past present problems to those who are 'othered' by the romantic ideals of the unified Irish nation: Protestants, Loyalists, immigrants, ethnic and linguistic others. Despite the occlusions produced by these streamlined histories, within the context of trauma and conflict they can be read as necessary for nation-building. 'Narrative, by constructing relationships between events (whether those events are empirically verifiable or not), gives them a relative value and a positional meaning. With narrative, the true becomes the real, even if that reality is recognized as a discursive formation.'[75] This repetitive historiographical pattern is reflected in Sands' work, producing the discursive logic of Irish Nationalism. As he states in a short essay 'Things Remain the Same – Torturous': 'There is no future in Ireland under oppression, only the same tragic history repeating itself in every decade.'[76]

The Manichean principles of Irish Nationalism are demonstrated in Sands' short prose piece 'The Window of My Mind,' written in Long Kesh:[77]

> On a dreary, dull, wet, morale-attacking November afternoon, when one's stomach is empty, and when the monotony begins to depress and demoralise, it is soothing in many respects to spend a half-an-hour with one's head pressed against the concrete slabs, gazing in wonder, and taking in the antics of a dozen or so young starlings bickering over a few stale crusts of bread. Circling, swooping, sizing up and daring an extra nibble, continually on their guard, and all their tiny nerves on end, the young starlings feud among themselves, the greedy one continually trying to dominate and always wanting the whole haul to himself, fighting with his comrades whilst the sparrow sneaks in to nibble at the spoils.
>
> But the ruler of the kingdom of my little twenty-yard arched view of the outside world, is the seagull,[78] who dominates, steals, pecks, and denies the smaller birds their share. The seagull takes it all. In fact, his appetite seems insatiable. He goes any length to gorge himself. Thus I dislike the seagull, and I often wonder why the starlings do not direct their attention to the predator, rather than each other. Perhaps this applies to more than birds.[79]

The metaphor is an obvious one, the final statement leaving little room for doubt that the scenario described by Sands is analogous to the warring factions in Northern Ireland. The seagull stands in for the ultimate colonizing threat: the English. This passage is important for several reasons. First and foremost, it anchors Sands' perceived struggle for power and agency in a symbolic order where food is the primary signifier. The piece even begins with an admission of Sands' own hunger, with the focus of the piece moving from the location of his own deprivation – his stomach – to those of the warring birds. Second, the description sets up a binary antagonism between the seagull – the English – and the starlings and sparrows, which represent the divided social community in Ireland: Protestants and Catholics, Nationalists and Loyalists, Republicans and neutral parties, etc.

Sands' insistence that the binary between the smaller birds and the seagull overrides the internal divisions among the starlings is reflective of his political belief that the sectarian conflict in Northern Ireland is reducible to an English–Irish dualism, despite the region's ongoing internal divides, which predate major British involvement during the Troubles. 'In so far as there is a consensus in the literature [regarding sectarian relations in Northern Ireland] then, it is that the picture is not black, nor white, but grey.'[80] During the Troubles, alliances were made and broken, separatist groups formed and fractured. Unionist versus Republican, Catholic versus Protestant: these groupings were not organized along stringent binaries that corresponded with an Irish–English divide.[81] Even the IRA itself separated into divided factions: the Real IRA, the Provisional IRA, the Official IRA, etc. The internal political landscape within Ireland was much too complicated to be reduced to a simplistic fight for liberation against the British. The role of Dublin, the position of the British government, the alliances between Unionists and the British – these relationships were extremely complex.[82] David McKittrick and David McVea present a comprehensive genealogy of the multiple warring factions over the course of the Troubles. Their depiction of the bloodiest years of the conflict demonstrates how complicated the political and social makeup of the region really was. But when they present an outsider's view of the internal turmoil in Northern Ireland, they state that:

> [M]any Irish-Americans simplistically viewed the conflict in Northern Ireland as a classic colonial struggle between British occupying forces and the gallant freedom-fighters of the IRA. In this romantic version of events many complicating factors, including the very existence of the Unionist population, simply did not exist when viewed from across the Atlantic.[83]

This simplistic view held by Irish-American sympathizers is presented as an inaccurate portrayal of the real politics of the region. Nevertheless, this transatlantic Nationalist ideology was deployed often and vigorously by the IRA who 'had seen itself as engaged in an anti-colonial, anti-imperialist freedom struggle.'[84] This was despite the fact that even nationalists such as John Hulme (founding member of Social Democratic and Labour Party)[85] and Garret FitzGerald (twice Taoiseach of the Republic of Ireland) had begun to evolve a political theory that largely disregarded the then-current role of the British:

> In this revised view the key to the problem was not Britain but the Protestant community. The import was that the British presence was not imperialist but neutral, that the border was maintained not because of British interests but at the insistence of the Unionists, and that Irish unity could only come about with Protestant consent. The real border, it was now said, was not geographic but in men's minds.[86]

Although this explanation has a Revisionist flavour, the political situation in Northern Ireland was and continues to be complex: 'The model of the troubles as a clash between two unreasonable warring tribes is thus a misleading or at least an incomplete picture.'[87] Nevertheless, the pervasive and durable nationalistic ideology that pitted British against Irish was reinvigorated and redeployed amid the growing sectarian unrest in the 1960s. The 15th Anniversary of the Easter Rising re-invoked older forms of Irish Republicanism. The IRA produced a teleological vision of history to fuel its nationalism and political action, and reached back to previous forms of Republicanism to structure an anti-colonial struggle that was seen as continuous and unfinished. 'Who started sectarianism? The English murderers who invaded Ireland.'[88] This was the fallback position of many IRA members, including Bobby Sands.

Sands reaffirms this in the aptly named prose piece 'We Won't be Fooled'. In it, he discusses the one true enemy to an unnamed cellmate:

> The Brits are up to every trick and there's nothing that would suit them better at the minute than a bloody sectarian campaign flaring up, as you know yourself comrade, the Brits are in a very bad way at the minute, because the IRA are inflicting many kills and casualties upon them, demoralising them and winning the war against them.[89]

In the previous image of the starlings and sparrows, the two types of birds are distinct but share common hardships, but the potential for alliances is undermined by the malevolent seagull. It is the same throughout Sands' writing. In his short story 'The Privileged Effort', Sands depicts a dialogue between himself and another unnamed cellmate:

> 'Hasn't it always been the same, hasn't there always been two sides – the privileged (them) and the oppressed (us). It's one half jailing the other, oppressing the other, murdering the other or whatever.'
> 'And we're always the other,' says the effort. 'Bloody brassnecks saying they're only doing a job.'[90]

Sands forces the past into the present, and the traditional binary between the British and the Irish is evoked to reproduce a history that is characterized by a long-standing mythical enmity and all the discursive power this evocation contains. The IRA campaign is legitimized and inserted into a narrative of a continuous, colonial struggle by Sands' own insistence and his investment in a teleology of the Irish struggle for self-determination.

The aims of the IRA owe much to Irish militant rather than cultural nationalisms. For Lloyd, the difference between cultural and militant nationalism lies mainly in their contrasting foundational myths. Lloyd maps this contrast through the figures of William Butler Yeats and Patrick Pearse. Yeats' vision of nation-building was based in a cultural unity as constructed through the literary recuperation of Ireland's past. In contrast, Pearse framed revolution through sacrificial violence that mythicized individual acts of bravery.[91] Ultimately, however, '[t]he martyr and not the poet was presented as the exemplary nationalist subjectivity.'[92] Consequently, this militaristic approach caused 'certain rhetorical consequences that issued in a failure to think explicitly about the shape of an independent Ireland. It would be enough simply to achieve independence.'[93] This single-minded pursuit of an independent Ireland is articulated through Sands' piece and accomplishes the dual task of strengthening morale in the fight against the British and urging sectarian factions in occupied Ireland to recognize the true threat that keeps the region from stability: the British. Nonetheless, these forms of nationalism were unable to address the numerous factions within Ireland that would require a far more complex solution for national co-existence than simply the dismantling of the Northern Irish union.

The way that metaphor constructs and mediates power is also represented in Sands' poem 'Modern Times', again through the symbol of food:

> In modern times little children die,
> They starve to death, but who dares ask why?
> And little girls without attire,
> Run screaming, napalmed, through the night fire.
> And while fat dictators sit upon their thrones,
> Young children bury their parents' bones,
> And secret police in the dead of night,
> Electrocute the naked women out of sight.[94]

This piece imagines an occupied, colonized people suffering from the wretched condition of starvation, an image that repeats often throughout Sands' prison writing. It correspondingly constructs the oppressive subject, the 'fat dictator,' as the cause of this unequal distribution of resources. The bodies presented here can be read through unfair economic conditions perpetuated by neoliberal markets. The theme of malnourished, lacking bodies contains within it a call for transformation; it demands a retributive justice that can be realized by the neutralization of extremes as embodied by excessive thinness and fatness. This poem refers to a contemporary context: 'It is said we live in modern times, In the civilized year of "seventy-nine."'[95] The historic context of the Troubles is clear. However, the imagery in the poem could very well be referring to Nationalist depictions of the Great Famine itself: children are burying their parents' bones, and the callous overlord gets fatter while the helpless subjugated waste away.[96] The title of the piece also draws attention to narratives of progress rooted in the rhetoric of Modernity that the conditions of Northern Ireland are failing to meet. In Northern Ireland, historical progress is short-circuited.

Instead, history is conceived as repetitive and rooted in the present; its failure to move on from the circumstances of the past gestures towards an underdevelopment of Modernity – which itself presents its own contradictions and problems. Sands' writing produces a normative teleology of history – an Enlightenment discourse of the forward march of civilization, traceable to British colonial thinking. For the English, the Irish always presented an inability to participate in its progressive ideals of Modernity. Sands highlights the backwardness of Ireland in order to inspire political action and provide a rationale for IRA Nationalisms. In 'Modern Times', he acknowledges that an inability to move on from the circumstances of the past is indicative of underdevelopment and barbarity:

> The difficulty faced by nationalists who would have used Ireland's destitution as a sign of difference is registered in the inability to step outside of a progressive historical metanarrative [...] Hence, a nationalist movement which was also a liberation movement was compelled to participate in the discourse of progress, even when that discourse was so deeply implicated in the project of imperialism.[97]

These sorts of contradictions – cementing an anti-colonialist struggle predicated on a unique Irish Nationalism, while still having to accede to the imperial rhetoric of Modernity in order to articulate these resistances – play out in the semantic and somatic field of Bobby Sands' hunger strike, as will be explored later in this chapter.

Food is the material object through which the determinations of power are articulated in Sands' writing. The possession of food is how bodies are codified as possessing or lacking power. Taking the food significations further still, in some of Sands' writing the IRA prisoner is metaphorically subject to the taboos of cannibalism[98]: he is killed, cooked, and eaten by an increasingly dehumanized British subject. In his prose piece 'Fenian Vermin, Etc.', Sands elaborates on the degrading, inhuman conditions of his imprisonment:

> The extent of this practice in H-Block would have to be heard and seen to be believed. Sort of makes you think why they keep us locked up and naked and filthy in disease-ridden, smelly concrete dungeon-like cells, unfit for pigs. It is a mentality that makes it so easy for them to torture us, or butcher us when the opportunity arises.[99]

The term 'butcher' may also allude to the infamous Loyalist faction of the Ulster Volunteer Force (UVF) that came to be known as the 'Shankill butchers.' This violent squad of Loyalists became notorious for their kidnapping, torture, and throat-slashing of Catholic civilians during the Troubles.[100] Again, the repetition of the word 'butcher' not only dehumanizes Catholic prisoners but also conflates Loyalist aims with the prison administration and the British. These descriptions refer to literal instances of physical torture, but on a symbolic level, the prisoner's subjectivity is under threat – threatened with engulfment by oppressive British forces. Another image of man-as-meat is mentioned in the poem in the second instalment of the 'Trilogy, Diplock Court':

> They walked me through the door of doom
> Like pig to slaughter pen.
> But pigs are treated better
> Than prisoners are, my friend.

> And I in lowly fetters
> Of captured Irishmen.[101]

The repetition of the pig imagery is a useful semiotic technique through which to deconstruct the Irish–English dichotomy, while it also serves to situate the nationalist struggles articulated by the IRA within a longer history of traumatic hunger. The metaphor of the pig does double duty in Sands' writing. As Sands writes in his lengthy poem, 'The Castlereagh Trilogy':

> I quaked like swine on the slaughter line
> As he said, "You're quite fit.
> So send him on, he's good and strong,
> And roast him on the spit."[102]

However, this imagery is complicated when Sands uses it in the opposing context of describing the British administration, particularly as it concerns the judge in the 'Diplock Court'[103] section of 'The Castlereagh Trilogy'. The judge is described, repeatedly, as 'pig-in-wig'[104] and 'fat pig':[105]

> The grunting pig he sneered and leered
> And scratched his lofty snout.
> He mumbled something rather snide
> That died as it crawled out,
> But carved a look upon his face
> That cast aside all doubt.[106]

The repeated use of the word 'pig' to describe the police and prison wardens adds to the split meaning of the term. A pig is both a degraded species in terms of abject filth and a mainstay of the European diet. A pig-as-prisoner is a helpless creature with no agency, but the pig-as-police represents an anomalous greed. Connotations of greed, consumption, the abject non-human, and the defenceless creature coalesce in the image of the pig.

This constellation of imageries aptly captures the politics and poetics of the history of Irish hunger, particularly national renditions that attribute the English government with explicitly necropolitical purpose. The starving Irish body – literally eating itself alive in starvation or being eaten by the greed of the British administration – is both an object of pity and abject horror. The genealogy of this nationalist explanation of hungry Irish bodies can be traced within Sands' writing and politics. The pig wants and eats more than it needs; the British were accused of sending much-needed stocks of food to its colonies and the imperial metropole during the famine. The pig is an animal that we consume with little forethought; the Irish population was subject to callous Malthusian economic policies that gave no thought to the humanity of the starving. The pig is coded as a particularly filthy and abject animal; this reflects both the inhumane conditions of Long Kesh prison and the abject state of bodies that starve to death (during famine, during hunger strikes) – both cause visceral disgust and are hard to look at, least of all imagine. The Irish have frequently been described as animal-like in famine writing, as Lloyd notes:

> It is no accident that [...] the animal is the most frequent figure for the Irish in their ambiguous location, for the spectator of famine seeks insistently to distance the famine victim from common humanity even as he or she extends sympathy.[107]

In both depictions of the Irish subject – in the prison and during the famine – body becomes less human and more object as it troubles the boundary between nature and culture. It is the subject of sympathy while also being distanced from the privileged subject of the human. Sands deploys the pig imagery as and when necessary, but, crucially, the image slips in and out of subject position – oppressor and oppressed. These tensions and doublings demonstrate how co-dependent the respective subject positions of oppressor and oppressed are. The contemporary socio-political context of Sands and his hunger is examined in closer detail in the following section, once again tracing the historical agon of Irish national identity as constructed against the British oppressor.

The 1981 Irish Hunger Strike: Contemporary Forms and Historical Contexts

The form of the hunger strike functions metonymically: the hunger striker's body is realized as the body politic, representative of a larger socio-cultural group and thus subject to the socio-political ideologies that govern the larger IRA/Republican mission statement, while the prison itself became a 'microcosm of Ireland.'[108] I read Sands' prison writing – which was expressly written with publication in mind, and so had a particular ideological register – for evidence of his political aims. I consider Sands' writing and instances of slippages or differences between what is perceived as the unifying IRA political aims of the strike and what may be perceived as Sands' own specific, individualized struggle for meanings on his own terms. These significations are certainly informed by the larger ideological pressures of the hunger strike as a political tool, but a derivative of those pressures creates a discrete struggle for agency, subjectivity, and power that is enacted by Sands himself, which is evidenced within his writing. I read the ambivalence found within his texts as proof of the contradictions inherent in the technique of starving as a tool for protest and agency, by the colonial/postcolonial subject. Sands' specific and challenging circumstances extend the original significations of the hunger strike to include a struggle to break free of constricting significations, while simultaneously revealing their boundedness.

The tradition of martyrdom and self-sacrifice in Ireland can be traced back to ancient Celtic culture and pagan Irish mythology. The fifth-century Brehon legal codes contained a precedent for fasting to obtain redress from an offender, or forgiveness for a debt, in the act of *troscad*. The aggrieved party, who generally had less power, wealth, and status than the creditor/accused, would publicly affirm the purpose of the ritual fast and do so at the door of the target, using the only tool at their disposal: the body. The basis of the fast was predicated on the open body models of pagan Ireland. The body of the starver not only presented a supernatural threat in the form of haunting if the starvation went unaddressed, but the logic of the permeable body reified the connection between the self and other – demanding redress buttressed by the social connections within the community. The social meanings of these Celtic ritual practices were later blended into Christian religious practices, thus adding an element of sacredness to fasting and combining with Catholic precepts of somatic self-control and sublimating the flesh and its vices. The tradition of fasting as redress against a more powerful individual continued in Ireland, focalized through Irish Nationalism and Republican struggle. Combined with cultural movements like the Gaelic literary revival and anti-colonial struggles such as the Easter Rising of 1916, the hunger strike became a well-defined strategy of resistance in Ireland. 'Between 1913 and 1922, more than 50 hunger strikes against the British administration were conducted by nearly 1,000 prisoners in different jails and internment camps.'[109] Popular Republican figures – such as Terence MacSwiney – became heroes for their strength of character and willingness to sacrifice for the greater cause of independence:

> The martyred dead were canonised in the popular consciousness and they were likened with the redemptive sacrifice of Christ, the martyrdom of saints, and to the heroes and rebels who had been dramatically instilled into the country. The sacrificial *motif* had been dramatically instilled into the Irish Catholic psyche, and in [Patrick] Pearse's terms, he and his fellow comrades had redeemed both their country and themselves.[110]

In an extension of this historic cult of self-sacrifice,[111] the hunger strike was a process in which the intended outcome was transformation. Through the somatic transformation of the hunger striker, the Irish nation itself would be transformed – along with the starvation's intended audience: British colonial forces. Hunger strikers must:

> persuade the people whom they fast against to take responsibility for their starvation. In this way, hunger strikers reveal the interdependency in which all subjects are enmeshed, because they force their antagonists to recognize that they are implicated in the hunger of their fellow human beings.[112]

The sorts of open/permeable body models of exchange of life for justice that structured the *troscad* were evoked in the 1981 hunger strikes but struggled against the competing logic of the closed, individuated post-Enlightenment, European body. Sands' writing attests to this disjunction – of the individual biopolitical body's struggle within the matrix of communal social representation.

The hunger strike 'disallow[s] such fluidity of body borders, when the oppressors try to invade them with food. They shut out the prison food from their bodies in the same way they try to shut out the British governance of Northern Ireland.'[113] These competing models of the *troscad* produce irresolvable contradictions in the hunger strike, symbolized through the body-as-nation. Ireland wishes to be an independent nation – closing off the borders, mimicked in the somatic borders of the hunger strike – but the *troscad* form relies on the open body model to compel the target, the British, to respond to the hunger strike – which they refused to recognize. The British policy of criminalization therefore presented a conflict of representational types and somatic-linguistic recognition. These contradictions play out in the 1981 hunger strike.

The hunger strike tapped into an ancient memory of popular uprising, to reconstruct it, and subsequently harness the resulting nationalist power. For the purposes of the IRA, the strike held wider material ramifications. In his study of the Irish hunger strikes, *Formations of Violence*, Alan Feldman notes the dual motives behind the hunger strike:

> The Blanketmen viewed the 1981 Hunger Strike as a military campaign and organized it as such. For them, it was a modality of insurrectionary violence in which they deployed their bodies as weapons. They fully expected a coupling of this act of self-directed violence with mass insurrectionary violence outside the prison. These two forms of violence were seen as semantically and ethically continuous.[114]

The anticipated publicity and consequent political action generated by the strike was the intended goal of the strike – again reasserting the continuous open body models of pagan and Catholic Irish discourse. It would disseminate the IRA cause far and wide, and indeed over the course of the strike individuals who were formally not involved with Republican or Nationalist agendas became politically active.[115] This wider goal of furthering the Republican movement was, no doubt, the primary logic behind the strikes, both for the prisoners who underwent them and the IRA army council outside.[116] A larger Catholic or Republican community was implicated symbolically into the fate of the prisoner's body: the body of the prisoner stood in for the collective, as body politic. 'The conversion of social consciousness was, as in other matters, founded on the conversion of the body and the self through ascetic disciplines.'[117] The ambivalences within Sands' writing and hunger complicate this easy reading of the body-as-community, situated as he was between the contradictions contained within Irish Nationalism, with its competing discourses of Enlightenment rationalities and romanticized Celtic nativism. If the hunger strikes attempted a 'closing off' of national borders through somatic enclosure, this was a performance of the idealized, individuated, closed subject – again, mirroring the European ideal of the nation state. However, this occludes the difference, liminality, and dislocations of those who inhabited Northern Ireland; the 'open' body of the *troscad* ideally would represent the whole community continuously. However, diverse factions of the community were necessarily elided by these stringent forms of nationalism as they were articulated through the hunger strike, as well as by the influences of the idealized individuated subjects produced by European Modernity which they may not have wished to conform to.

The hunger strikers' five demands[118] essentially amounted to the right to be legitimized as political prisoners, constituting, in particular, a response to the 'criminalization' policy that was being undertaken by the British government.[119] The strike was less about achieving the five demands and more about what the demands signified – an accepted politicization of IRA violence as legitimately recognized warfare, and the recognition that the Troubles were directly linked to an ancient conflict between the British and Irish. Bobby Sands restates this. He writes in the first entry of his hunger strike journal:

> I am dying not just to attempt to end the barbarity of the H-Block, or to gain the rightful recognition of a political prisoner, but primarily because what is lost here is lost for the Republic and those wretched oppressed whom I am deeply proud to know as the "risen people."[120]

The potential to improve prison conditions was secondary to gaining the right to signify the prisoner's body and inserting this legitimized political body into a specific, polarized historical narrative that portrayed the British as antagonist and the Irish as self-determining subjects. As Feldman notes: 'The Hunger Strike, state interrogation, and terrorist violence are practices united by the assumption that the inscription of violence on the human body (protracted or sudden violence) constitutes the production and display of truth.'[121] The British administration and the IRA were fighting over what can be counted as – in its visible signs on the body and what is accepted ideologically – true. Thatcher's government insisted on branding it as a criminal body; Thatcher herself famously stated in a televised interview, 'Crime is crime is crime: it is not political, it is crime.'[122] Her repetition of the word "crime" is clearly a semiotic strategy – she recognizes the constitutive power of language. The site of contested meaning was in this instance the prisoner's body. The hunger strike became a conflict of representation.

Stripping away prison clothes during the 'blanket protest' that preceded the 1981 hunger strike substantiates the initial logic of the hunger strike. Maud Ellman states: 'the spectacle of nakedness titillates the clothed with the delusion of their own superiority.'[123] The blanket protest inverts the politics of degradation in its struggle to redefine the body and interrogates the limits of somatic representation while also motioning toward its constructed nature. In this formulation, the blanket men's nakedness is not only a meaning-making intervention into the Irish body, but again requires a co-option of the British observer to structure its logic. As the site of intersection between Britain's criminalization policy and a competing Irish Republican ideology of freedom fighters, the prisoner's body was not merely the site of conflict that connected them; it was, in fact, constructed *by* these competing representational frictions.

By deconstructing the relationship between the two competing categories of body and representation, and the equally competing categories of British identity as aggressor and Irish identity as freedom fighter – and recognizing the interned Irish body cannot be divorced from these interlocking discourses – it becomes clear why the hunger strike's logic is fraught with contradiction.

Although Sands was not on the hunger strike for most of the time spent in prison, nor while he wrote most of his prison works, I utilize Sands' entire prison oeuvre as a basis for textual analysis. The deployment of a hunger strike as a means of protest was always an available option, both because of the long-standing tradition of hunger striking in Irish Republicanism and due to the specific context of Long Kesh prison at the time. The 1981 hunger strike closely followed the initial failed hunger strike in 1980, which demonstrates that prisoners never considered the failure of the first strike a deterrent to striking again; the possibility of self-starvation was a continued presence in the prison even when no one was officially striking. Second, it is clear from Sands' prison writing that hunger was a prosaic part of internment. 'I cursed the cold, my aching body and the pangs of hunger that never left me.'[124] Sands constantly refers to hunger in his writing, particularly as a tool to break prisoners' spirits and resistance. 'Darkness and intense cold, an empty stomach and the four screaming walls of a filthy nightmare-filled tomb to remind me of my plight, that's what lay ahead tomorrow for hundreds of naked Republican Political Prisoners-of-War.'[125] Along with the physical isolation of the cell and the conditions of internment, hunger enforced by the prison itself is interpreted as a form of torture. The four walls of his cell become an extension of Sands' body.[126] Sands is at the mercy of the prison 'screws' (slang for prison wardens) survey and control – through the biopolitical control of imprisonment and of the body and its biological needs. In the context of this understanding of hunger, Sands' hunger strike is seen as the solution to a particular set of problems contained within the experience of internment and the various socio-political meanings attached to it.

Body of Text: Bobby Sands and the Hunger Strike

The 1981 IRA hunger strike was a somatic response to imprisonment – literal and biopolitical. The prison strike's purpose is first and foremost a strategy to raise awareness of the political struggle. The striking body produces speech forms when language cannot; the hunger strike is the speech of the voiceless. When the prisoners undertook the blanket protest – they were denying the British administration, represented by the prison guards, the right to signify their body with prisoners' uniforms. But while the prisoners challenged what meanings were heaped on to the surface of their bodies, they also resisted what went *into* their bodies. This intrusion took two forms. The first was the forcible searches that took place on a regular basis – prisoners were anally probed for contraband in a ritualized humiliating practice. Representations of this were scant in prison writing due to its particularly emasculating nature, although Sands does mention it briefly in his journal/novella *One Day in My Life*:

> I dropped my towel, turned a full circle and stood there embarrassed and naked, all eyes scrutinizing my body.
> 'You forgot something,' the mouthpiece grunted.
> 'No I didn't,' I stammered in a fit of bravado.
> 'Bend down tramp,' he hissed right into my face in a voice that hinted of a strained patience. Here it comes, I thought.
> 'I'm not bending,' I said.
> Roars of forced laughter reinforced by a barrage of jibes and abuse erupted.
> 'Not bending!' the confident bastard jibed.
> 'Not bending! Ha! Ha! He's not bending, lads,' he said to the impatient audience.
> Jesus, here it comes. He stepped beside me, still laughing and hit me. Within a few seconds, in the midst of the white flashes, I fell to the floor as blows rained upon me from every conceivable angle [...] Someone had my head pulled back by the hair while some pervert began probing and poking my anus.[127]

The scene establishes the prison administration's policy of intense surveillance of and into prisoners' bodies; the prison's biopolitical gaze is literally signified and metonymically represents the perceived invasion of sovereign Irish land by hostile British forces.

> The colonizer remains in Northern Ireland, and, especially during the troubles of the 1970s and 1980s, conducts its business on the bodies of Irish people through combat, crossing their body boundaries while artificially bolstering the created boundary between North and South.[128]

Both the surveillance and cavity search are technologies of power. These forms of somatic control were met with resistance. The practical reasons for invasive bodily searching are secondary to its ability to assert power over the imprisoned body: 'They were not interested in searching me but in humiliating me!'[129] This particular passage also establishes the association between 'bending' or 'breaking' as tantamount to breaking into the body of the prisoner, thus interfering with the prisoner's subjectivity, and also producing a set of anxieties around the masculinity and the coherency of the male, heterosexual body.

The second way in which prisoners' bodies were given meaning was through the intake of food. In Long Kesh, the allocation of food was a contentious issue between screws and inmates. Here the prisoners were presented with a difficult double bind. They had to eat to survive; to continue utilizing their bodies as a site of protest. They needed to remain alive 'to fight back with all we really have.'[130] However, the intake of food itself was a form of capitulation. The meagre diet provided to prisoners practically illustrates the screws' power over their bodies and, furthermore, in the act of eating the prisoners were placed in a position wherein they had to symbolically and literally swallow food that symbolized British ideology and colonial force. 'Taken that the psychological dimension of eating [...] is often embedded in suppressed emotions of anger and denial, swallowing becomes a subversive metaphor of acceptance and refusal of what is unfair.'[131] In this forcible power play, the prisoners are at the mercy of the biological and political needs of their bodies:

> Funny how one can adapt to things – especially when you are starving, I thought, remembering the times during the summer when the orderlies and screws had dropped maggots into our dinners and all we could do was search for them and remove them, then eat our dinners as if nothing had happened. It was either that or starve!'[132]

Sands often describes the terrible condition of prison food in his writing. In a passage in *One Day in My Life*, he describes how he is intentionally served mouldy bread and manages to spot and discard it just before consuming it. The decaying bread is a tell-tale symbol of the prison administration's lack of respect for the prisoner's body and is read as contamination yet cannot be refused. Every act of ingestion is a sign of humiliation and another instance of British oppression. 'Disgusted, I literally forced the meagre bit of bread and lukewarm tea into me.'[133] In this instance, it is the prisoner himself that is forced to render his body abject with the polluting symbols of British occupation.

Considering the prisoners' reaction to the treatment of their bodies – the forced anal searches, the distasteful prison food on offer – it is clear that the prisoners considered the screws' actions an ideological as well as literal invasion, thus rendering their bodies colonized in much the same fashion they considered Ireland occupied by foreign invaders. Sands makes sure to embed his current struggle in a continuous history of British domination and occupation:

> But have things not always been the same for Irish Republican Prisoners of War incarcerated in British hell-holes? [...] The repetition continues as the present generation of Irish men and women likewise rot and die and are relentlessly tortured, and the next generation, and the following generations may prepare to meet the same fate unless the perennial oppressor – Britain – is removed for she will unashamedly and mercilessly continue to maintain her occupation and economic exploitation of Ireland to judgement day, if she is not halted and ejected.[134]

Likewise, the prisoners' aim was to disable and deny the screws' invasion into their bodies by closing off the exits. This is a literal interpretation of the hunger strike. 'The boundaries of their homeland and now the boundaries of their identity are consistently broken, and they begin to use body boundaries to protest.'[135] By regulating the physical boundaries of their subjectivity through the denial of food into their bodies, the prisoners' hunger strike was an attempt to shore up the borders of their occupied bodies and nation.

Like the other hunger strikers examined in this book, Sands uses a Cartesian understanding of the body to articulate his hunger protest. This involves a division between the material body and the discursive realm of the 'mind,' 'spirit' or 'soul.' Foucault cautions that although the latter is a representational category, it is an integral component of the Cartesian dualism, and is anything but illusory:

> [I]t would be wrong to say that the soul is an illusion or an ideological effect. It really does exist, being produced perpetually: it is produced permanently around, on, within the body of the functioning of power that is exercised on those punished – and, in a more general way, on those one supervises, trains and corrects, over madmen, children at home and at school, the colonized, over those who are stuck at a machine and supervised for the rest of their lives.[136]

Sands' motivation for deploying this oppositional binary is a logical rejoinder to the discipline and control his body is subject to in the prison environment. As his body is under siege, he retreats into the realm of the mental, where a limitless, substance-less freedom is located and imagined. 'I have but one weapon to overcome them: my own thoughts.'[137] His mind is idealized as a space that cannot be colonized by invading forces, and this division is repeated again and again. 'My body wants to say: "Yes, yes, do what you will with me. I am beaten, you have beaten me." I am beaten, you have beaten me. But my spirit prevails.'[138] In his writing, the spirit is allied with flight, liberation, and the image of the skylark – which represents freedom. The flighty imagery is diametrically opposed to the burdensome weight of the body. In his poem 'The Rhyme of Time', Sands conceives of a universal spirit of man, unattached and yearning for the God-given right of freedom:

> It lies in the hearts of heroes dead,
> It screams in tyrant's eyes,
> It has reached the peak of mountains high,
> It comes searing 'cross the skies.
> It lights the dark of this prison cell,
> It thunders forth its might,
> It is 'the undauntable thought,' my friend,
> That thought that says 'I'm right!'[139]

The category of thought, or spirit, is valorized, eternal, and flies, in contrast to the material realities of the prison, which are represented as enclosure and suppression. It is in the elevated language of representation that the material world can be devalued as an unimportant space. The mind, the spirit – these are spaces that the screws, the British, cannot access or interfere with.

Invariably, the dichotomy between the mind and the body is rendered interdependent, both on a literal level and a discursive one. The hunger strike itself utilizes the body as a bargaining chip, and to even enact such a protest, the body must remain in play. It is the only weapon the prisoners have in their possession. Indeed, although Sands attempts to disappear into this airy category of the spirit – a safe and controllable location – his descriptive language in relaying this category slips dangerously close to how he describes the screws' attempted invasion, or 'breaking', into his body. From his prose piece 'Alone and Condemned': 'God, life is hard for the oppressed, but to fight back is a victory. To remain unbroken in spirit is a great victory.'[140] The concept of 'breaking' into the spirit mirrors the attempted breaking into the prisoners' bodies by the prison administration:

> My spirit cries out, arise, but my body pleads for no more, for mercy, and it wants to lie upon the cold black blood-splattered ground and die, to sleep, for it is weary and broken and perhaps dying each minute, each eternal minute.[141]

A deconstructive reading reveals the internal problematic of using the mind/spirit/representational to escape the realities of bodily harm. The use of language to contain and articulate a linguistic escape from physical pain reveals additional problems, as Elaine Scarry reminds us: 'Physical pain does not simply resist language but actively destroys it.'[142] The physicality of the body undoes the hunger strike's logic of privileging the representational and linguistic.

Nevertheless, Sands utilized writing as a means of escape from the trials of the body. Writing and language provided an arena to exercise some agency and gave shape and meaning to his imprisonment and fractured subjectivity, which he clearly delineates in his work as a soldier in a war for freedom from oppression. This is perhaps why he produced such a wealth of writing in prison, and repeatedly – almost obsessively – reiterated his position in a legitimate war of liberation.

> I am a political prisoner. I am a political prisoner because I am a casualty of a perennial war that is being fought between the oppressed Irish people and an alien, oppressive, unwanted regime that refuses to withdraw from our land.[143]

Sands formulates this identity in this manner repeatedly in his writing, as though the act of writing itself could redefine his body. Ellmann notes how often a period of food abnegation is accompanied by occasions of intense production or consumption of texts. 'Since reading and writing mime the processes of eating and excreting they provide a kind of methadone for the obsession.'[144] As the process of eating prison food was viewed as enforced ingestion and colonial oppression, the hunger strikers chose to reject food, and the material needs of the body, but writing became a way of feeding the soul/mind while the body was under siege.

Even until the last recorded day of his hunger strike, Sands read voraciously.[145] He read and received papers, comms from other prisoners, and letters from friends, family, and supporters outside Long Kesh:

> Got papers and a book today. The book was Kipling's *Short Stories* with an introduction of some length by W. Somerset Maugham. I look an instant dislike to the latter on reading his comment on the Irish people during Kipling's prime as a writer: 'It is true that the Irish were making a nuisance of themselves.' Damned too bad, I thought, and bigger the pity it wasn't a bigger nuisance!¹⁴⁶

Sands defines what 'correct' sorts of words should be ingested into the body. He remains on a steady diet of political and creative writings that confirm his Nationalist and Republican ideology. This is articulated through the learning, use, and teaching of Gaelic. Prisoners both used Gaelic to shield communications between one another from the prison guards and used the language to emphasize an idealized pre-colonial Irish Nationalism:

> The Gaelic kept the whole thing together. The education programs traced the history of Ireland right back to the Ice Age. How Ireland as part of the European continent separated and right the way down through this history, the Viking invasions, the British invasions right down to the development of Republican Ideology.¹⁴⁷

Unlike Nyasha in the previous chapter, Sands signals his distrust of the literature of the colonial masters, and their potential to colonize the mind – to establish in Eagleton's words, a cultural hegemony. Nyasha's strategy to shore up the troubled boundaries of her identity and body is to ingest the colonial discourse and starve out the native body. Sands' strategy may seem more foolproof; he solidifies the borders of his body with the rejection of the colonizer's food, discourse, language, and somatic interference – exemplified in this hunger strike:

> Gaelic inextricably tied to the mobility and transcendence of the disembodied voice, the solidarity of collective vocality, as well as deep historical resonances, overcame the semiotics of captivity. The acquisition of Gaelic, with all its multiple uses and manifold social meanings, functioned as a mechanism of decontamination.¹⁴⁸

These forms of decontamination are problematized nonetheless by their investment in a romantic ideal of pre-colonial Ireland, one that again taps into narratives of Irish history that elide historical complexity, national traumas (such as the famine and colonial rule), and heterogeneous Irish communities. There are contradictions inherent in his strike, just as they are in Nyasha's, Nimi's, and Michael K's – all arising from a complicated colonial identity and the internal contradictions of the hunger-striking mechanism as focalized through a dichotomized nationalism.

Sands' anxiety around defining the body and its meanings resulted from the fractious subjectivity of the colonial subject. This expressed itself through the

Manichean structures of Irish Nationalism as previously explored, but there are several other troubled states that constitute the colonial subject in Ireland. The Irish body contained multiple internal struggles and contradictions. The strict British/Irish division that structured Irish national identity did not prevent the sorts of internalized inferiority complexes that define the imperial subject, a 'deep sense of cultural inferiority of the kind so ably documented in other colonial contexts.'[149] Added to the psychosocial dislocations of the colonial subject, there is considerable evidence that 'that Irish people have particular psychic and character vulnerabilities that emerge from a pervasive sense of inferiority and malignant shame exacerbated by the effects of an Irish Catholic upbringing steeped in puritanical morality.'[150] 'The internalisation of cultural negative self-images meant that "self-hating" rhetoric was certainly a recurring and important feature of Irish nationalism'[151] because 'Irish nationalists accepted many contemporary anti-Irish stereotypes.'[152]

The Irish colonial body is racially codified in distinct ways. Since the nineteenth century, the Irish race has occupied a liminal space. At the time, the concept of race was an unstable one and was undergoing the process of scientific rationalization that produced a hierarchy of races. '"Race" could be synonymous with "nation" – the distinction was seldom clear; or it could apply to a family of nations, notably the Celts, who together were said to compose a single race.'[153] British notions of the Celt's backwardness compared them to the '"savages" of North America and sometimes to the "Hottentots" of South Africa, who were commonly seen as the "lowest" of the savage races' – they were '"wild", "indolent" and "superstitious."'[154] These kinds of racial qualifications sit uneasily within contemporary racial models that are more predicated on skin colour, but:

> [i]n the eyes of wealthy Protestants who claimed the right to construct hierarchies of race and nationality – in England and Ireland as well as the United States – the Catholic Irish were the bearers of cultural characteristics that made them like blacks. They too were perceived as indolent, irrational, childlike. Some observers speculated that their historical roots were racially ambiguous or even African; others compared them to gorillas and chimpanzees.[155]

Thus the Irish occupied a liminal racial space:

> constructed as neither 'white' colonisers nor 'black' colonised, but were instead imagined to inhabit a discomfiting zone of racial indeterminacy [...] This idea of racial indeterminacy, of being neither white nor black [...] had an important impact on Irish identity-formation. It was a major cause of the push by Irish nationalists to create a more idealised and stable self-image.[156]

Thus, it 'is it not entirely comprehensible, then, that the Irish would have needed to claim the mantle of whiteness for themselves'[157] in order to legitimize their claim to self-sovereignty? Although these forms of racialization of the Irish body had all but disappeared by Sands' time, they had equally been replaced with a form of cultural racism that upheld much of the same logic:

> These caricatures suggested that the Irish were less evolved than their English neighbours, less advanced along a set path of historical development. While such perceptions may not have outlived the nineteenth century, they would cast a long shadow over Irish nationalist thought.[158]

These forms of racial inferiority produced in the Irish consciousness types of self-hatred that were common to colonial and postcolonial subjects. Irish Nationalism and its aims of independence were complicatedly tied up with this racial discomfort. 'Irish nationalism was a concerted effort to disprove such stereotypes and create a more prideful self-image of a "white" nation. Crafting an image of strong and racially redeemed Irish men was a key part of this.'[159] The difficulty with these forms of nationhood is that they uncomfortably intersected with racist colonial discourse. If the Irish were seen as unstable, violent, backward, and out of step with Modernity – and therefore unable to self-govern – Irish Nationalism sought to upend these stereotypes that they themselves had internalized by investing in a European model of state-building:

> The form and content of Irish nationalism, how it conceived of the state and of the economy, as well its tropes of muscular men and subtly expressed anxieties about racial purity, were not only a reaction against British rule but also at the same time, paradoxically, strongly influenced by contemporary British politics.[160]

This troubled and simultaneously reinforced Irish Nationalism's oppositional construction of itself against Britain. By attempting to use British discourses of Modernity to their own ends, Irish nationalists became the imitators of their oppressors, and, in fact, justified racist stereotypes of the Irish as being incapable of self-rule. If they rejected the models of rationality produced by the metropole, they would retrench themselves within the barbarism of the British caricatures of the Irish.

Simultaneously, the Irish body was feminized in ways that complicated the muscular forms of Irish Nationalism that sought to subvert, but that also reproduced the colonial order in contradictory ways:

> Irish passivity in the face of British intrusion only reinforced colonialist notions about the inherent passive femininity of Irish men, whilst violent resistance was decried as irrationality, a prominent sign that childish Irish men still needed paternal British supervision. Representations of masculinity in Irish culture and politics were [...] an attempt to break free of this bind and disprove these negative tropes of Irish men's passive femininity or violent irrationality.[161]

Ireland's relationship with England is often viewed in maternalistic terms; the Irish are 'cast in the role of dependent child. Implicit in this child-colony/mother-country relationship was the threat that the 'child' would one day grow up and seek independence and separation.'[162] Within the context of Sands' contemporary politics, Irish Republicanism responded to these anxieties about the femininized Irish body by producing a robust, hypermasculine IRA leadership as an ideal. Tamar Mayer argues that even when the nation is viewed as feminine, nationalism itself is viewed as a masculine behaviour.[163] These readings of 'Mother Ireland' in need of protection by the strong, self-sacrificing masculine freedom fighter reinforced a gendering of Irish Republicanism. During the Troubles, the IRA focused their discursive attack on England upon the feminine figure of Margaret Thatcher. Emasculated by the British state as focalized by a woman, Irish Republicans sought to defend an already historically imperilled masculinity with the controlled 'rational' violence of the hunger strike. The idealized masculine IRA man also explains why anxieties produced by the strip searches – the boundaries of the masculine IRA body forcibly broken by the British state forces – were interpreted along gendered lines and contained a potent shame. The ideal Republican is hypermasculine – most valorized in martyrdom and presented as 'instructive'[164] to other Irish men, and their forms of militaristic intervention were interpreted as the most anti-colonial.[165] Here again, the traumatic narrative of Irish Nationalist struggle (a Manichean, militaristic struggle against colonial forces) is redeployed and circulates in the political matrix of Ireland, synchronizing past and present. The militant nationalisms used to form the hunger strike can be traced to the ongoing – yet atemporal – binaries produced by the trauma of Irish history. These forms of masculine nationalities had the effect of 'collapsing of historical time served to create a highly ideological, atemporal zone wherein all true Irish nationalist men exist together.'[166] As Sands himself writes: '[u]nfortunately, the years, the decades, and centuries, have not seen an end to Republican resistance in English hell-holes, because the struggle in the prison goes hand-in-hand with the continuous freedom struggle in Ireland.'[167] The IRA hunger strike can be read as an oversimplistic yet extreme solution to a complicated contemporary, historically informed, socio-political context.

In the context of these competing and contradictory histories and discourses, the hunger strike can be read as a re-enactment of the biopolitical control and discipline that operates on the macro-level of Northern Ireland, the micro-level of the prison, and finally on the individualized level of the body itself. To counter the irrationality associated with the feminized, racially ambiguous, and backward Irish body, Sands deploys a Cartesian logic of biopolitical control on himself. Using the rationality of the mind – associated with the rational Modernities espoused by the colonial state – Sands tries to legitimize his political body by demonstrating mental self-control. This cult of self-sacrifice dovetails with the long history of Irish Nationalism inflected by a Catholic practice of the mortification of the flesh. Republicans put their bodies on the line to perform a masculinity through the somatic technique of hunger. This hunger, however, had to demonstrate control and suppress bodily instinct and desire. In this way, the Irish hunger striker sought to refute the irrationality heaped on the Irish body, and essentially 'starved out' the feminine, racialized, and animalistic stereotypes that historically structured Irish subjectivity. This self-styled violence therefore sought to distance itself from barbaric violence that the British state signified as criminal and savagely terroristic. On the contrary, the form of the hunger strike is predicated on rational self-restraint. This performance of self-control was an attempt to subvert the spatial control of the prison by eluding the bodily terrors contained within – by escaping to the realm of the logical, 'free' mind. Through the starvation of Sands' body, the Catholic/Republican community is also transformed and 'rationalized,' thus more in line with the Enlightenment ideals of European Modernity. This situated Sands' hunger at the intersections of several competing discourses that emerge from Irish Republicanism and forms of Nationalism that are predicated on a masculine, militaristic, eternal struggle against the British.

The starving Irish body, historically produced by the traumas of colonialism and famine, is situated within a series of interlocking dichotomies. Caught in the simplified Manichean struggle with British imperial forces (traced in Irish famine historiography, and the history of Irish Nationalism and Republicanism) and the gendered and racialized profiles produced by colonial discourse – the starving Irish body is caught in a series of paradoxes. The hunger strike is an attempted intervention into these contradictory positions, using a Cartesian model of the body. The forms of the hunger strike themselves contain within them the seeds of their own undoing. The sublimating of the Irish body – and all its associated negative connotations – by the rational/logical mind only reaffirms the logic of metropole discourses of development, progress, and Modernity. Irish Nationalism faces particularly difficult routes through this version of nation-building, because it presents a valid form of statehood as enshrined in the British model of Modernity – which Ireland both wanted to participate in and was also weary of as this would erase the differences between Ireland and Britain that propelled the Irish national project forward. Similarly, by suppressing the racial ambiguity contained in the Irish body, Ireland colluded in the racist discourses of empirical Enlightenment thought – and tried to ally itself with those forces of racial inequality that categorized those other colonized peoples that the Irish identified with. By espousing an exclusive masculine, self-sacrificial model of Irish Republicanism, IRA men faced another threat. The hunger strike was already problematized by this summation of Sands' identity: were the hunger strike resolved as planned, symbolizing and metonymizing British capitulation, the Republican identity Sands valued – and which he insistently constructed as essential – would be in jeopardy. 'But I'll just lie here and continue to resist them knowing that some day our day will come [...] Me? I'll always remain the same – an Irishman fighting for the freedom of my oppressed people.' To sublimate the body according to the needs of the rational mind requires an erasure of the self, and paradoxical reaffirmation of those colonial discourses that encode the Irish body as a degraded subaltern figure. As Ellman states: 'a hunger strike, which, far from sabotaging the idea of order, legitimates the very powers that hold it ransom.'[168]

By situating the starving prison striker within the historical context of famine trauma, however, we may recuperate and extract some value from its contradictions, and its contested status as a successful anti-colonial strategy. By linking Sands' starving body to those from the Great Famine and reading the connection through the traumatic narratives the famine produced, we can untangle some productive modes of thinking through Sands' hunger. If trauma has the effect of 'engineering its reappearance through symptoms,' by contextualizing Sands' starvation through the national trauma of the Great Famine, we might read his hunger as a re-articulation of traumatic and repressed hunger, and an attempted resolution through forms of repetition. This may seem like a huge historical leap to make, and certainly the historical contexts between the two temporal geographies – Northern Ireland in 1981 and Ireland in 1845 – require an attentiveness to historical oversimplifications. However, as is evidenced by my analysis above, there are several points of contact between the two instances of historical hungers. 'The principal slogan of the 1981 hunger strike evoked the unity of past and present.'[169] The continuity of Irish Nationalisms and historiography, Ireland's constant devotion to hunger techniques, and a familiar anti-colonial response to European imperial discourses (both historical and contemporary, articulated through the civilized/savage binary) produce a genealogy of Irish history, nationalism, and the body that links the two events together. Irish Nationalism's approach to European Modernity, in particular, helps to bridge the historical gap.

> The sanctification of the idea of progress in the mid-nineteenth century has a direct bearing on the writing of the Famine. Like a regal, theocratic juggernaut, the metanarrative of steady, inevitable human improvement pushed all other structures for ordering events in time beyond the discursive pale into the realm of the barbarian.[170]

Post-Enlightenment ideals of progress are engaged in explicit ways in Sands' hunger strike technique. In order to 'progress' out of the realm of the barbarian, Sands rationalized his body using the Cartesian model of the self. This privileges the progressive ideals of the logical mind over the irrational, feminized, racially othered body. This technique aimed to reinsert Sands' starving body into the progressive narratives of Modernity, which were interpreted as the ideal route toward nation-building. The famine is read as a temporal event that not only excludes Ireland from the historical narratives of progress and Modernity but is necessarily done so to reify the supremacy of European modes of rationality. 'In other words, progress inevitably marches on, not in spite of famine, pestilence, and other forms of "transient evil" – but because of them, under the sign of "Supreme wisdom."'[171] Thus, we can see the traces of colonial discourse within Sands' strike, and the invariable contradictions they produced from which he struggled to break free.

But if we read Sands' hunger as a reassertion of a traumatic history – inscribed on his starving body – we might think of ways of reading his hunger as a recuperation of a traumatic history that is elided by both Nationalist and Revisionist histories, national silences, and robust, masculine IRA national identity. Through the repetition of hunger, Sands asserts agency over the historically occluded famine body: 'the hunger striker has a chance to exert a certain control over the communication process, which she has set in motion with her fast.'[172] Sands uses his body as a somatic speech form that displays the abject horror of starvation. If language is not up to the task of representing trauma, then perhaps Sands' body – in all of its somatic violence – can speak of a past too painful for words to contain. 'He devoids his body of the fat that represents its frozen past.'[173] He 'makes flesh' the suffering of the starving Irish body, but by self-selecting this form, he reroutes the politics of sympathy that codifies the famine victim – relegated into the oblivion of silent un-representability – into a prominent display of self-imposed starvation. The discourse of the starving body is displayed in the hieroglyphs of his diminishing body. By transforming the position of victim-of-hunger to agent-of-hunger, Sands re-articulates a history of food insecurity that exists in the affective zone of the body. His body rewrites a history of the famine that moves beyond the stringent binaries of Nationalist and Revisionist history – it individuates, it traumatically repeats, and it displays the real pain of starvation. Of course, Sands would himself no doubt insert his body into a Nationalist tradition of British/Irish antagonism. But if we deploy a writerly approach to the body, rather than the body Sands himself purports to narrate, we can find productive ways of negotiating Sands' hunger.

Kelleher argues that depictions of the Great Famine have been feminized, particularly in famine writing. This refers to literal representations of the famine, whereby effects of starvation are often illustrated through the female body, 'where the spectacle of a hunger body is created, this occurs, predominantly, through images of women.'[174] The tendency to feminize also gestures to the traumatic limits of representation, and the role of the feminine in constructions of the unspeakable – that which exists beyond the rationality of language: again, the body/mind dualism reappears in these formulations, with the embodied feminine category representing the irrationality that exceeds the rational realm of discourse. 'In the specific context of famine literature, one encounters both writers' sense of "unspeakable," that which is too awful to relate, and also [...] their attempts to give it form; in this regard, female images emerge.'[175] The female famine victim possesses an allegorical function, embodying the fate of Mother Nature of the feminized nation itself. Women become the texts on to which the trauma of the famine can be transferred. Rather than presenting as sources of testimony, they become symbols in a larger national narrative strategy that may 'cover over the "unspoken", that which needs to be spoken, to be remembered and retold.'[176] The feminization of famine does double duty – it articulates a gendered narrative of trauma that positions the feminine as the limits of rational representation, and it also protects the male starving body from vulnerability by erasing representations of it. This, of course, suits the purposes of a robust Irish nationalism based on hypermasculinity.

However, Sands reinserts the individuated starving male body into the larger historical image of the Great Famine. This individuated hunger not only responds to calls from historians to think through famine historiography from a more nuanced position than the Nationalist/Revisionist battle, but it also transfers the responsibility of women as the 'bearers of meaning' to the male form. This somatic repetition of starvation thus exceeds the unimaginable trauma of the Great Famine and seeks an alternative form to articulate the unspeakable. Although Sands' hunger strike was certainly a display of masculine control over the somatic, we may be able to read alternative meanings in his somatic speech:

> The performance of protest suicide as well as the premise on which it is based enlist rationality – exchange and value calculation; strategies for attracting notice, resisting authority, and swaying public opinion – although imagining fills in at critical turns [...] Imagining converts metaphors into reality and vice versa, frequently employing mimesis as catalyst.[177]

What Karin Andriolo refers to here as 'imagining' is deployment of affective communication – our own imagining:

> Whereas rational thought stays within the mind, imagining does not; it tentacles into the body. When we imagine an object or a scene, our senses get involved [...] Imagination grabs mind and body. What is felt in one's body speaks a mightier authenticity than an abstract idea in one's brain. Imagining is embodied minding.[178]

It is the affective register produced by Sands' starvation that allows for a reading that transcends his intended meanings, to be transmuted into something more radical and productive within the context of representation. The starving male Irish figure is a form of testimony for a larger historical narrative of traumatic hunger, interwoven with the traumas of colonialism, spoken in a form whose affective speech form might more accurately and radically represent the occluded truth of Irish history. The immense amount of national grief produced by Sands' death formalizes the act of mourning that may have been short-circuited in the traumatic silence of the famine.

After ten hunger strikers died in Long Kesh prison, the families of the remaining men on strike began to perform medical interventions to revive them once they fell into the tell-tale coma that preceded imminent death. The British government conceded to most of the demands but did not grant the much-fought-over right for designation as special category prisoners. They would not accept defeat in the discursive war fought through the prisoners' somatic protest. In the end, they remained, to the British, terrorists and not prisoners of war. However, social commentators argue that Sands' death was not in vain. His election as a Member of Parliament[179] while on hunger strike demonstrated the power of mainstream political interventions to the IRA and Sinn Féin. Sinn Féin became much more active in mainstream Northern Irish politics after the strike,[180] participating in brokering the Good Friday Agreement that signalled the end of the Troubles. The prisoners brought worldwide attention to their cause, shamed their oppressors, and captured the sympathies of their community. As an anti-colonial strategy, it certainly produced some measure of success. However, when we examine Sands' writing closely, we see that the prison environment heightened, and thus altered, the biopolitical control of prisoners, separating them from forms of control experienced by Northern Irish Catholics outside:

> precisely because they are produced within a special matrix – at once removed from the bounds of everyday society, yet also within the undiluted heart of that society as replicated in its disciplinary structure – prison texts are able to cast light upon subjects quite external to the physical prison cell, subjects that may in fact be invisible to those outside.[181]

The prison was a space where the discourses of power were magnified, substantiated by the violence and discipline of the imprisoned human body. Sands associates this surveillance and bodily torture allegorically to the violence of the state, although the unremitting and extreme form of his prison experience undoubtedly produced consequences that far exceeded the conditions of subjugation that characterized life outside. For Sands, there existed two distinct motivations for going on the hunger strike: there were the overtly political motivations as set out by the IRA, but also a suggestion that death became a legitimate possibility of escape from prison. 'We balance precariously on the thin divide between sanity and insanity, every aspect of our existence is cloaked in torture.'[182] The sense conveyed by Sands' writing is that he is a man on the brink, unable to take much more of the torture and isolation of prison. The hunger strike is realized as the desire to overcome and resolve these unbearable thresholds. He no longer wishes to be the 'living dead.'[183] He wishes to be either dead or alive, and the prison is portrayed as a type of purgatory where the only guarantee is a half-lived existence: 'I wish I were dead. "But I am dead," I say aloud: I can't even kill myself, I think.'[184] The desire for an ending – to the continual torturous passage of time, to isolation, to the enforced beatings and searches – speaks to the problematic of using a Cartesian model of protest fast within the context of extreme pain, of attempting to escape the body and its pain through language.

> We have fought for freedom and we still fight for freedom with all we have left, our only weapon our spirit, but in our nakedness the spirit cannot repel the wolf or shield the blow of the baton or deflect the pulverising raining punches, it cannot repel torture![185]

Scarry states that torture reduces the world 'to a single room or set of rooms.'[186] This problematizes the allegorical reach of Sands' protest – if not for his audience, then certainly to himself. 'In the torture rooms of Northern Ireland, all such traces [of the human] were erased; even language was replaced [...] broken only by the sound of mortal cries.'[187] The discursive breaks down at the point of extreme pain, so for Sands and the other prisoners in Long Kesh, the prison transcended the Nationalist ideology used to animate the protest themselves. We witness Sands' own growing unease with Cartesianism in his final hunger strike journal entry:

> I was thinking today about the hunger-strike. People say a lot about the body, but don't trust it.
>
> I consider that there is a kind of fight indeed. Firstly the body doesn't accept the lack of food, and it suffers from the temptation of food, and from other aspects which gnaw at it perpetually.
>
> But the body fights back sure enough, but at the end of the day everything returns to the primary consideration, that is, the mind. The mind is the most important.
>
> But then where does this proper mentality come stem from? Perhaps from one's own desires for freedom. It isn't certain that that's where it comes from.[188]

Sands loosens his grip on the belief that the Cartesian model can transform the self, body politic, and nation. There seems to be a dissolution between the category of mind/body, a deconstructive turn that illustrates their interdependence, and an admission that the body's mortality – in wasting away – possesses greater power of expression than Sands initially anticipated. But if we read the somatic speech forms of the body, we can read beyond the purported purposes of the strike, and perhaps glean a greater understanding of both the traumatic history of Ireland and the inculcations of imperialism and their legacies. This applied reading of the somatic provides its own anti-colonial potential, even if its occurrence is situated far from both the original context of the hunger strike and the intentions of the starving body itself.

Notes

1. The start of the Troubles is difficult to pinpoint, having begun with isolated incidences of violence that gradually escalated into a more serious and widespread form of regional violence. It is accepted that they started in the 1960s and concluded with the signing of the Good Friday Agreement in 1998.
2. James Donnelly, Jr., *The Great Irish Potato Famine* (Stroud: Sutton Publishing, 2001), p. 3.
3. Ibid., p. 8.
4. See Cormac Ó Gráda, *Ireland Before and After the Famine: Explorations in Economic History, 1800–1925* (Manchester and New York: Manchester University Press, 1993), pp. 2–8.
5. Austin Bourke, *The Visitation of God? The Potato and the Great Irish Famine*, ed. by Jacqueline Hill and Cormac Ó Gráda (Dublin: Lilliput Press, 1993), p. 52.
6. Donnelly, p. 9.
7. Cecil Woodham-Smith, *The Great Hunger: Ireland 1845–9* (London: Hamish Hamilton, 1962), p. 30.

8. Joel Mokyr, *Why Ireland Starved: A Quantitative and Analytical History of the Irish Economy, 1800–1850* (London: Allen & Unwin, 1985).
9. Cormac Ó Gráda and Phelim P. Boyle, 'Fertility Trends, Excess Mortality and the Great Irish Famine', *Demography*, 23.4 (1986), 542–563.
10. Ó Gráda is less convinced by theories of a collective memory, suggesting that it elides individual experience. He asks, 'What, one may wonder, is the life span of a collective trauma?' (114) and argues that 'one of the points at issue here is that the famine did *not* inflict "common injuries" at the same time, never mind across generations.' My reading of the famine is one of collective trauma that lingers in the public consciousness and unconsciousness, intergenerationally, in unquantifiable and shifting ways. I take a less literal interpretation of trauma than Ó Gráda: Ó Gráda, Cormac, 'Famine, Trauma and Memory', *Béaloideas*, 69, (2001), 121–143 (p.141).
11. Ó Gráda, Ireland Before and After the Famine, p. 98.
12. Judith Butler, *Undoing Gender* (London: Routledge, 2004), p. 153.
13. Raphaël Ingelbien, 'Elizabeth Gaskell's "The Poor Clare" and the Irish Famine', Irish University Review, 40.2 (2010), 1–19 (p. 2). The quotation in the last sentence is from: Niall O'Ciosáin, 'Was there "Silence" about the Famine?' *Irish Studies Review*, 13 (Winter 1995/96), 7–10, (p.7); Frank O'Connor's article 'Murder Unlimited' appeared in The Irish Times, 10 November 1962, p.8.
14. Margaret Kelleher, *The Feminization of Famine: Expressions of the Inexpressible?* (Durham, NC: Duke University Press, 1997), p. 2.
15. Dori Laud and Nanette C. Auerhahn, 'Knowing and Not Knowing Massive Psychic Trauma: Forms of Traumatic Memory', *International Journal of Psychoanalysis*, 74 (1993), 287–302 (p. 288).
16. David Lloyd, 'The Indigent Sublime: Specters of Irish Hunger', *Representations*, 92.1 (2005), 152–185 (p. 162).
17. Donnelly, p. 14.
18. See John Mitchel, *The Last Conquest of Ireland (Perhaps)*, ed. by Patrick Maume (Dublin: University College Dublin Press, 2005).
19. Colm Tóibín, 'The Irish Famine', in *The Irish Famine: A Documentary*, ed. by Colm Tóibín and Diarmaid Ferriter (London: Profile Books, 1999), pp. 28–29.
20. For a more detailed analysis of revisionist and post-revisionist accounts of Irish history, see Declan Kiberd, *Inventing Ireland: The Literature of The Modern Nation* (London: Jonathan Cape, 1995); F.S.L. Lyons, *Ireland Since the Famine* (London: Fontana, 1985); Roy Forster, *Modern Ireland 1600–1972* (London: Penguin, 1989); Tom Dunne, *Rebellions: Memoir, Memories and 1798* (Dublin: Lilliput Press, 2004); and Paul Bew, *The Politics of Enmity: 1789–2006* (Oxford: Oxford University Press, 2007).
21. Michael O'Loughlin, 'Trauma Trails from Ireland's Great Hunger: A Psychoanalytic Inquiry', in *Loneliness and Longing: Conscious and Unconscious Aspects*, ed. by Brent Willock, Lori C. Bohm, and Rebecca Coleman Curtis (Hove: Routledge, 2012), 233–250 (p. 237).
22. Donnelly, p. 13.
23. Ó Gráda, *Ireland Before and After the Famine*, p. 99.
24. See Ciarán Ó Murchadha, *The Great Famine: Ireland's Agony, 1845–1852* (London: Bloomsbury Academic, 2013); Enda Delaney, *The Great Irish Famine: A History in Four Lives* (Dublin: Gill & Macmillan, 2014); and John Kelly, *The Graves Are Walking: A History of the Great Irish Famine* (London: Faber and Faber, 2012).
25. Donnelly, p. 215. See also Peter Solar, *Famine: The Irish Experience 900–1900*, in *Subsistence Crises and Famines in Ireland*, ed. by Margaret E. Crawford (Edinburgh: John Donald Publishers, 1989), 112–131.
26. See Breandán Mac Suibhne, *Subjects Lacking Words? The Grey Zone of the Great Famine* (Hamden, CT: Quinnipiac University Press, 2017).

27 The 'grey zone' of the Great Famine included accounts of murder over food and resources, euthanizing children and cannibalism. These harrowing accounts are often excluded from more didactic representations of the Great Famine.
28 Breandán Mac Suibhne, 'A Jig in the Poorhouse', *Dublin Review of Books*, 32 (2013), https://drb.ie/articles/a-jig-in-the-poorhouse [accessed 13 October 2018].
29 Breandán Mac Suibhne, 'A Jig in the Poorhouse'.
30 It is worth noting, however, that there are also problems with using individual testimony as representational entry points into conceptualizing the totality of famine. The sheer scale of death produced by famine can be effaced by these kinds of individual accounts.
31 David Lloyd, *Irish Times: Temporalities of Modernity* (Dublin: Field Day, 2008), p. 34.
32 Michelle Balav, 'Trends in Literary Trauma Theory', *Mosaic: An Interdisciplinary Critical Journal*, 41.2 (2008), 149–166 (p. 149).
33 Ibid., p. 150.
34 Kali Tai, *Worlds of Hurt: Reading the Literature of Trauma* (New York: Cambridge University Press, 1996).
35 Cathy Caruth, *Trauma: Explorations in Memory* (Baltimore, MD: Johns Hopkins University Press, 1995), p. 152.
36 Roger Luckhurst, *The Trauma Question* (London: Routledge, 2008), p. 8.
37 G. Kearns, '"Educate That Holy Hatred": Place, Trauma and Identity in the Irish Nationalism of John Mitchel', *Political Geography*, 20 (2001), 885–911 (p. 888).
38 See Tim Pat Coogan, *The Famine Plot: England's Role in Ireland's Greatest Tragedy* (New York: Palgrave Macmillan, 2012).
39 Terry Eagleton, 'Form and Ideology in the Anglo-Irish Novel', *Bullán*, 1.1 (1994), 12–26 (p. 17).
40 Lloyd, *Irish Times*, p. 65.
41 Kearns, p. 891.
42 Ingelbien, p. 2.
43 Lloyd, *Irish Times*, p. 64
44 Astrid Erll, 'Literature, Film, and the Mediality of Cultural Memory', in *Cultural Memory Studies: An International and Interdisciplinary Handbook*, ed. by Astrid Erll and Ansgar Nünning (Berlin: Walter De Gruyter, 2008), 389–398 (p. 392).
45 Katherina Dodou, 'Jim Sheridan's "The Field" and the Memory of Dispossessed Irishness', *Nordic Irish Studies*, 13.1 (2014), 111–128 (p. 114).
46 Chris Morash, *Writing the Irish Famine* (Oxford: Clarendon, 1995), p. 4.
47 Sean Ryder, 'Reading Lessons: Famine as the Nation, 1845–1849', in *Fearful Realities: New Perspectives on the Famine*, ed. by Chris Morash and Richard Hayes (Dublin: Irish Academic Press, 1996), 151–163 (pp. 160–161).
48 For further discussion on the silence around the Irish Famine, see Niall O Cioséin, 'Was There "Silence" About the Famine?', *Irish Studies Review*, 4.13 (1995), 7–10.
49 Lloyd, *Irish Times*, pp. 43–44.
50 O'Loughlin, p. 137.
51 Ibid., p. 242.
52 Patrick Brantlinger, 'The Famine', *Victorian Literature and Culture*, 32.1 (2004), 193–204 (p. 203).
53 Anne Whitehead, *Trauma Fiction* (Edinburgh: Edinburgh University Press, 2004), p. 3.
54 O'Loughlin, p. 242.
55 No doubt, however, that the impact of the capitalist colonialism that replaced and overtook traditional Irish political economy during the famine plays out in Ireland's present-day contexts, both culturally and economically.
56 Ó Gráda, 'Famine, Trauma and Memory', pp. 140–141.
57 Kearns, p. 900.

58 In his memoir, Hugh Dorian frames the famine inside a narrative of colonial oppression, although he also (somewhat contradictorily) notes: 'Were we permitted to moralise we would say that the Almighty in His wise ways has brought about dispersions, emigrations and deaths as a punishment upon the people, as they were too numerous, too unruly, and in their ways of life [...] too rebellious; therefore, a Higher Power was needed to curb and to chastise them': Hugh Dorian, *The Outer Edge of Ulster: A Memoir of Social Life in Nineteenth-Century Donegal*, ed. by Breandán Mac Suibhne and David Dickson (Dublin: Lilliput Press, 2000), p. 172.

This particularly Malthusian explanation reminds us that two explanations of the famine – cruel political economy, cruel providence – have more in common than they may initially seem to do. For a deeper analysis of the intersections of Malthus and Adam Smith, see Chapter 1.

59 Christine Kinealy, 'Beyond Revisionism: Reassessing the Great Irish Famine', *History Ireland*, 3.4, (1995), 28–34 (p. 28).

60 Terry Eagleton, *Heathcliff and the Great Hunger: Studies in Irish Culture* (London and New York: Verso, 1995), p. 29.

61 See H.V. Brasted, 'Irish Nationalism and the British Empire in the Late Nineteenth Century', In *Irish Culture and Nationalism 1750–1950*, ed. by Oliver MacDonagh, W.F. Mandle and Pauric Travers (London: Palgrave Macmillan, 1983).

62 Woodham-Smith., p. 19.

63 Kiberd, p. 9.

64 Patrick O'Farrell, *England and Ireland Since 1800* (London: Oxford University Press, 1975), p. 4.

65 Kearns, p. 889.

66 Benita Parry, *Postcolonial Studies: A Materialist Critique* (London: Routledge, 2004), p. 65.

67 Lloyd, *Irish Times*, p. 64.

68 Carole Blair, Greg Dickinson and Brian L. Ott, 'Rhetoric/Memory/Place', in *Place of Public Memory: The Rhetoric of Museums and Memorials*, ed. by Carole Blair, Greg Dickinson and Brian L. Ott (Tuscaloosa: University of Alabama Press, 2010), 1–56 (p. 10).

69 Although it is important to note that Bobby Sands' Republicanism was strongly affiliated with socialism, the IRA's Republicanism still subscribed to post-Enlightenment ideals of Modernity and statehood.

70 Morash, p. 54.

71 Richard English, *Irish Freedom: A History of Nationalism* (Basingstoke: Macmillan, 2006), p. 26.

72 Kearns, p. 895.

73 Although it is worth noting that Michel did later move away from this sort of community-building strategy. See Kearns.

74 English, p. 377.

75 Morash, pp. 13–14.

76 Bobby Sands, *Skylark Sing Your Lonely Song: An Anthology of the Writings of Bobby Sands* (Dublin and Cork: Mercier Press, 1982), p. 136.

77 Sands spent four and a half years in the H-Blocks of Long Kesh. Most of his writings were written on scraps of paper and toilet paper that were smuggled out of the prison over these four years. Except for the diary Sands kept during the first 21 days of his hunger strike, the specific dates for the majority of these writings are unknown. Sands did not date them, was interned during their publication, and did not survive to provide dates upon publication of the anthology used in this chapter. Sands was interned in 1977 and died in 1981. The writings were composed over this four-year period.

78 *Faolean*, the Irish word for seagull, was also the Gaelic term Sands and other prisoners used to describe the prison guards in Long Kesh.

79 Sands, *Skylark*, p. 86.
80 J.H. Whyte, *Interpreting Northern Ireland* (Oxford: Oxford University Press, 1990). p. 168.
81 This is also due to direct English intervention in the 1970s, in the context of growing sectarian violence in Northern Ireland.
82 See David McKittrick and David McVea, *Making Sense of the Troubles* (London: Penguin Group, 2000).
83 Ibid., p. 126.
84 Ibid., p. 167.
85 The Social Democratic and Labour Party (SDLP) is the main nationalist party in Northern Ireland, established in 1970 with the aim of promoting a united Ireland by nonviolent means.
86 Ibid., p. 165.
87 Ibid., p. 238.
88 Anonymous, 'Sectarianism!', *Republican News*, 19 May 1973, p. 1.
89 Sands, *Skylark*, p. 35.
90 Ibid., p. 109.
91 David Lloyd, *Anomalous States: Irish Writing and the Postcolonial Moment* (Dublin: Lilliput Press, 1993).
92 Kearns, p. 902.
93 Ibid., p. 896.
94 Sands, *Skylark*, p. 18.
95 Ibid.
96 Bobby Sands writes directly of the events of the Great Famine in several pieces, the most obvious being 'Ghosts in My Tomb', a poem that references mass emigration, starvation, and the 'coffin ship' *The Star of Hope*. The language used in this poem is also similar to the language in 'Modern Times,' with the British colonizer directly referenced in the second stanza: 'And the English Lord on the pheasant fed and dined and wined and grew fat' (*Skylark*, p. 120 and p. 18).
97 Morash, p. 58.
98 This imagery of cannibalism may also allude to the repressed 'grey zone' of representations of the Great Famine, see footnote 28.
99 Sands, *Skylark*, p. 29.
100 McKittrick and McVea, p. 115.
101 Sands, *Skylark*, p. 59.
102 Ibid., p. 47.
103 The Diplock Courts were a judiciary system established by the British government to speed up the process of Republican prosecution during the Troubles. It was juryless and presided over by a single judge. These courts were abolished in 2007.
104 Sands, *Skylark*, p. 61.
105 Ibid., p. 60.
106 Ibid., p. 61.
107 Lloyd, *Irish Times*, p. 51.
108 Karin Andriolo, 'The Twice-Killed: Imagining Protest Suicide', *American Anthropologist*, 108.1 (2006), 100–113 (p. 103).
109 Ibid., p. 103.
110 George Sweeney, 'Self-Immolative Martyrdom: Explaining the Irish Hungerstrike Tradition', *Studies: An Irish Quarterly Review*, 93.371 (2004), 337–348 (p. 339).
111 For more on the concept of surrogation, see Joseph Roach, *Cities of the Dead: Circum-Atlantic Performance* (New York: Columbia University Press, 1996).
112 Ellmann, Maud, *The Hunger Artists: Starving, Writing and Imprisonment* (London: Virago Press, 1993), p. 54.

113 Miriam O'Kane Mara, 'Food, Hunger, and Irish Identity: Self-Starvation in Colum McCann's "Hunger Strike"', in *Food and Literature*, ed. by Gitanjali G. Shahani (2018), 319–334 (p. 325).
114 Allen Feldman, *Formations of Violence: The Narrative of the Body and the Political Terror in Northern Ireland* (Chicago, IL: University of Chicago Press, 1991), p. 220.
115 Ibid., p. 220.
116 It is important to note that at any given time there were more IRA members in prison than outside. This created the unique condition of interned IRA members wielding significant political power within the ranks of the entire organization, and arguably more so the IRA council – with whom they disagreed on several issues, including the 1981 hunger strike. Thus, there was a discontinuous narrative of political aims within the IRA itself, emphasizing the dislocated and intolerable state of internment.
117 Ibid., p. 226.
118 (1) The right to wear civilian clothes; (2) the right not to engage in prison work; (3) the right to free association with other prisoners; (4) 50 per cent remission of their sentence; and (5) normal visiting schedules, parcels, and recreational and educational facilities.
119 'Criminalization' or 'Ulsterization' was a security and political policy deployed during the Troubles that meant that IRA and other paramilitary groups were to be labelled and treated as criminals and denied any legitimate political motivations. This was part of a strategy that discouraged the perspective of paramilitary groups from participating in a war, and instead encouraged a view of them as common thugs or criminals.
120 Sands, *Skylark*, p. 153.
121 Feldman, p. 226.
122 Radio One, 'Margaret Thatcher Press Conference Ending Visit to Saudi Arabia', BBC Radio News Report, 21 April 1981.
123 Ellmann, p. 102.
124 Bobby Sands, *One Day in My Life* (Cork: Mercier Press, 2001), p. 32.
125 Sands, *One Day*, p. 117.
126 The dirty protest, whereby prisoners slopped human waste on the walls of their cells, cements this logic of body-as-prison and prison-as-body. The dirty protest is another iteration of the abjectification of the self, a category that the hunger strike also operates within.
127 Sands, *One Day*, p. 28.
128 Mara, p. 320.
129 Sands, *One Day*, p. 70.
130 Sands, *Skylark*, p. 135.
131 Lin Elinor Pettersson, 'Neo-Victorian Incest Trauma and the Fasting Body in Emma Donoghue's *The Wonder*', *Nordic Irish Studies*, 16 (2017), 1–20 (p.14).
132 Sands, *One Day*, p. 92.
133 Ibid., p. 31.
134 Ibid., p. 136.
135 Mara, p. 235.
136 Michel Foucault, *Discipline and Punish: The Birth of the Prison* (New York: Vintage, 1995), p. 29.
137 Sands, *One Day*, p. 82.
138 Ibid., p. 90.
139 Ibid., p. 111.
140 Ibid., p. 93.
141 Ibid., p. 128.
142 Elaine Scarry, *The Body in Pain: The Making and Unmaking of the World* (Oxford: Oxford University Press, 1985), p. 4.

143 Sands, *One Day*, p. 153.
144 Ellmann, p. 24.
145 Sands eventually stopped writing the account of his hunger strike on day 21, to preserve his energy for the duration of the strike.
146 Sands, *Skylark*, p. 156.
147 Feldman, p. 312.
148 Feldman, p. 216.
149 O'Loughlin, p. 235.
150 Ibid., p. 235.
151 Aidan Beatty, *Masculinity and Power in Irish Nationalism, 1884–1938* (London: Palgrave Macmillan, 2016), p. 36.
152 Ibid., p. 34.
153 Bruce Nelson, *Irish Nationalists and the Making of the Irish Race* (Princeton, NJ: Princeton University Press, 2012), p. 6.
154 Ibid., p. 6
155 Ibid., p. 9
156 Beatty, p. 5.
157 Nelson, p. 19.
158 Beatty, p. 36.
159 Ibid., p. 4.
160 Beatty, p. 4.
161 Ibid., p.10
162 Thomas Bartlett, 'Theobald Wolfe Tone: An Eighteenth-Century Republican and Separatist', *The Republic: A Journal of Contemporary and Historical Debate*, 2 (2001) 38–46 (p. 43).
163 Tamer Mayer, *Gender Ironies of Nationalism: Sexing the Nation* (New York: Routledge, 2000).
164 Blair, Dickinson, and Ott, p. 10.
165 Colin Graham, '"Blame it on Maureen O'Hara": Ireland and the Trope of Authenticity', *Cultural Studies*, 15.1 (2001), 58–75 (p. 60).
166 Beatty, p. 22.
167 Sands, *Skylark*, p. 163.
168 Ellmann, p. 21.
169 Andriolo, p. 103.
170 Kelleher, p. 16.
171 Ibid., p. 23.
172 Andriolo, p. 103.
173 Ellmann, p. 88.
174 Kelleher, p. 8.
175 Ibid., p. 7.
176 Ibid., p. 7.
177 Andriolo, p. 108–109.
178 Ibid., p. 101.
179 Sands was nominated as parliamentary candidate for Fermanagh and South Tyrone under the label 'Anti H-Block/Armagh Political Prisoner', and narrowly won the seat. He held the post as an absentee.
180 McKittrick and McVea, p. 143.
181 Lachlan Whalen, *Contemporary Irish Republican Prison Writing: Writing and Resistance* (New York: Palgrave Macmillan, 2007), p. 6.
182 McKittrick and McVea, p. 137.
183 Ibid., p. 72.
184 Ibid., p. 104.
185 Ibid., p. 127.
186 Scarry, p. 40.

187 Ellmann, p. 100.
188 McKittrick and McVea, p. 173.

Bibliography

Andriolo, Karin, 'The Twice-Killed: Imagining Protest Suicide', *American Anthropologist*, 108. 1 (2006), 100–113

Anonymous, 'Sectarianism!', *Republican News*, 19 May 1973, p. 1

Balav, Michelle, 'Trends in Literary Trauma Theory', *Mosaic: An Interdisciplinary Critical Journal*, 41. 2 (2008), 149–166

Bartlett, Thomas, 'Theobald Wolfe Tone: An Eighteenth-Century Republican and Separatist', *The Republic: A Journal of Contemporary and Historical Debate*, 2 (2001), 38–46

BBC Radio One, 'Margaret Thatcher Press Conference Ending Visit to Saudi Arabia', BBC Radio News Report, 21 April 1981

Beatty, Aidan, *Masculinity and Power in Irish Nationalism, 1884–1938* (London: Palgrave Macmillan, 2016)

Bew, Paul, *The Politics of Enmity: 1789–2006* (Oxford: Oxford University Press, 2007)

Blair, Carole, Dickinson, Greg and Ott, Brian L., 'Rhetoric/Memory/Place', in *Place of Public Memory: The Rhetoric of Museums and Memorials*, ed. by Carole Blair, Greg Dickinson and Brian L. Ott (Tuscaloosa: University of Alabama Press, 2010), 1–56

Bourke, Austin, *The Visitation of God? The Potato and the Great Irish Famine*, ed. by Jacqueline Hill and Cormac Ó Gráda (Dublin: Lilliput Press, 1993)

Brantlinger, Patrick, 'The Famine', *Victorian Literature and Culture*, 32. 1 (2004), 193–204

Brasted, H.V., 'Irish Nationalism and the British Empire in the Late Nineteenth Century', in *Irish Culture and Nationalism 1750–1950*, ed. by Oliver MacDonagh, W.F. Mandle and Pauric Travers (London: Palgrave Macmillan, 1983)

Butler, Judith, *Undoing Gender* (London: Routledge, 2004)

Caruth, Cathy, *Trauma: Explorations in Memory* (Baltimore, MD: Johns Hopkins University Press, 1995)

Coogan, Tim Pat, *The Famine Plot: England's Role in Ireland's Greatest Tragedy* (New York: Palgrave Macmillan, 2012)

Delaney, Enda, *The Great Irish Famine: A History in Four Lives* (Dublin: Gill & Macmillan, 2014)

Dodou, Katherina, 'Jim Sheridan's "The Field" and the Memory of Dispossessed Irishness', *Nordic Irish Studies*, 13. 1 (2014), 111–128

Donnelly, Jr., James, *The Great Irish Potato Famine* (Stroud: Sutton Publishing, 2001)

Dorian, Hugh, *The Outer Edge of Ulster: A Memoir of Social Life in Nineteenth-Century Donegal*, ed. by Breandán Mac Suibhne and David Dickson (Dublin: Lilliput Press, 2000)

Dunne, Tom, *Rebellions: Memoir, Memories and 1798* (Dublin: Lilliput Press, 2004)

Eagleton, Terry, 'Form and Ideology in the Anglo-Irish Novel', *Bullán*, 1. 1 (1994), 12–26

Eagleton, Terry, *Heathcliff and the Great Hunger: Studies in Irish Culture* (London and New York: Verso, 1995)

Ellmann, Maud, *The Hunger Artists: Starving, Writing and Imprisonment* (London: Virago Press, 1993)

English. Richard, *Irish Freedom: A History of Nationalism* (Basingstoke: Macmillan, 2006)

Erll, Astrid, 'Literature, Film, and the Mediality of Cultural Memory', in *Cultural Memory Studies: An International and Interdisciplinary Handbook*, ed. by Astrid Erll and Ansgar Nünning (Berlin: Walter De Gruyer, 2008), 389–398

Feldman, Allen, *Formations of Violence: The Narrative of the Body and the Political Terror in Northern Ireland* (Chicago, IL: University of Chicago Press, 1991)

Forster, Roy, *Modern Ireland 1600–1972* (London: Penguin, 1989)

Foucault, Michel, *Discipline and Punish: The Birth of the Prison*(New York: Vintage, 1995)

Graham, Colin, '"Blame it on Maureen O'Hara": Ireland and the Trope of Authenticity', *Cultural Studies*, 15. 1 (2001), 58–75

Ingelbien, Raphaël, 'Elizabeth Gaskell's "The Poor Clare" and the Irish Famine', *Irish University Review*, 40. 2 (2010), 1–19

Kearns, G., '"Educate That Holy Hatred": Place, Trauma and Identity in the Irish Nationalism of John Mitchel', *Political Geography*, 20 (2001), 885–911

Kelleher, Margaret, *The Feminization of Famine: Expressions of the Inexpressible?* (Durham, NC: Duke University Press, 1997)

Kelly, John, *The Graves Are Walking: A History of the Great Irish Famine* (London: Faber and Faber, 2012)

Kiberd, Declan, *Inventing Ireland: The Literature of the Modern Nation* (London: Jonathan Cape, 1995)

Kinealy, Christine, 'Beyond Revisionism: Reassessing the Great Irish Famine', *History Ireland*, 3. 4, (1995), 28–34

Laud, Dori and Auerhahn, Nanette C., 'Knowing and Not Knowing Massive Psychic Trauma: Forms of Traumatic Memory', *International Journal of Psychoanalysis*, 74 (1993), 287–302

Lloyd, David, *Anomalous States: Irish Writing and the Postcolonial Moment* (Dublin: Lilliput Press, 1993)

Lloyd, David, *Irish Times: Temporalities of Modernity* (Dublin: Field Day, 2008)

Lloyd, David, 'The Indigent Sublime: Specters of Irish Hunger', *Representations*, 92. 1 (2005), 152–185

Luckhurst, Roger, *The Trauma Question* (London: Routledge, 2008)

Lyons, F.S.L., *Ireland Since the Famine* (London: Fontana, 1985)

Mac Suibhne, Breandán *Subjects Lacking Words? The Grey Zone of the Great Famine*, (Hamden, CT: Quinnipiac University Press, 2017)

Mac Suibhne, Breandán, 'A Jig in the Poorhouse', *Dublin Review of Books*, 32 (2013), www.drb.ie/essays/a-jig-in-the-poorhouse [accessed 13 October 2018]

Mara, Miriam O'Kane, 'Food, Hunger, and Irish Identity: Self-Starvation in Colum McCann's "Hunger Strike"' in *Food and Literature*, ed. by Gitanjali G. Shahani (2018), 319–334

Mayer, Tamer, *Gender Ironies of Nationalism: Sexing the Nation* (New York: Routledge, 2000)

McKittrick, David and McVea, David, *Making Sense of the Troubles* (London: Penguin Group, 2000)

Mitchel, John, *The Last Conquest of Ireland (Perhaps)*, ed. by Patrick Maume (Dublin: University College Dublin Press, 2005)

Mokyr, Joel, *Why Ireland Starved: A Quantitative and Analytical History of the Irish Economy, 1800–1850* (London: Allen & Unwin, 1985)

Morash, Chris, *Writing the Irish Famine* (Oxford: Clarendon, 1995)

Nelson, Bruce, *Irish Nationalists and the Making of the Irish Race* (Princeton, NJ: Princeton University Press, 2012)

O Cioséin, Niall, 'Was There "Silence" About the Famine?', *Irish Studies Review*, 4. 13, (1995)

Ó Gráda, Cormac, 'Famine, Trauma and Memory', *Béaloideas*, 69 (2001), 121–143

Ó Gráda, Cormac, *Ireland Before and After the Famine: Explorations in Economic History, 1800–1925* (Manchester and New York: Manchester University Press, 1993)

Ó Gráda, Cormac, and Boyle, Phelim P., 'Fertility Trends, Excess Mortality and the Great Irish Famine', *Demography*, 23. 4 (1986), 542–563

Ó Murchadha, Ciarán, *The Great Famine: Ireland's Agony, 1845–1852* (London: Bloomsbury Academic, 2013)

O'Farrell, Patrick, *England and Ireland Since 1800* (London: Oxford University Press, 1975)

O'Loughlin, Michael, 'Trauma Trails from Ireland's Great Hunger: A Psychoanalytic Inquiry', in *Loneliness and Longing: Conscious and Unconscious Aspects*, ed. by Brent Willock, Lori C. Bohm, and Rebecca Coleman Curtis (Hove: Routledge, 2012), 233–250

Parry, Benita, *Postcolonial Studies: A Materialist Critique* (London: Routledge, 2004)

Pettersson, Lin Elinor, 'Neo-Victorian Incest Trauma and the Fasting Body in Emma Donoghue's The Wonder', *Nordic Irish Studies*, 16 (2017), 1–20

Roach, Joseph, *Cities of the Dead: Circum-Atlantic Performance* (New York: Columbia University Press, 1996)

Ryder, Sean, 'Reading Lessons: Famine as the Nation, 1845–1849', in *Fearful Realities: New Perspectives on the Famine*, ed. by Chris Morash and Richard Hayes (Dublin: Irish Academic Press, 1996), 151–163

Sands, Bobby, *One Day in My Life* (Cork: Mercier Press, 2001)

Sands, Bobby, *Skylark Sing Your Lonely Song: An Anthology of the Writings of Bobby Sands* (Dublin and Cork: Mercier Press, 1982)

Scarry, Elaine, *The Body in Pain: The Making and Unmaking of the World* (Oxford: Oxford University Press, 1985)

Solar, Peter, *Famine: The Irish Experience 900–1900*, in *Subsistence Crises and Famines in Ireland*, ed. by Margaret E. Crawford (Edinburgh: John Donald Publishers, 1989)

Sweeney, George, 'Self-Immolative Martyrdom: Explaining the Irish Hungerstrike Tradition', *Studies: An Irish Quarterly Review*, 93:371 (2004), 337–348

Tai, Kali, *Worlds of Hurt: Reading the Literature of Trauma* (New York: Cambridge University Press, 1996)

Tóibín, Colm, 'The Irish Famine', in *The Irish Famine: A Documentary*, ed. by Colm Tóibín and Diarmaid Ferriter (London: Profile Books, 1999), 28–29

Whalen, Lachlan, *Contemporary Irish Republican Prison Writing: Writing and Resistance* (New York: Palgrave Macmillan, 2007)

Whitehead, Anne, *Trauma Fiction* (Edinburgh: Edinburgh University Press, 2004)

Whyte, J.H., *Interpreting Northern Ireland* (Oxford: Oxford University Press, 1990)

Woodham-Smith, Cecil, *The Great Hunger: Ireland 1845–9* (London: Hamish Hamilton, 1962)

Index

Page numbers followed by 'n' refer to notes.

affect theory 14, 93, 95
African Communist 69, 71
alimentary discipline and food 43
alimentary rituals, communal identity expression through 39, 44
Anderson, Judith 73
Andrade, Susan Z. 112
Andriolo, Karin 167
anorexia 102, 111–114, 119, 125n2
anorexia protests: against colonial racism 114; against gender inequality 112–114; *see also Nervous Conditions*
Arnold, David 46. 47
Attridge, Derek 70, 90, 91
Atwell, David 71, 92–93

Bagchi, Amiya Kumar 53
Bahri, Deepika 1, 110, 112
Barker, Clare 110
Barthes, Roland 7, 24
Beeton, Isabella 19, 42
Belizean culture, and food habits 10
Berlant, Laurent 14
Bethlehem, Louise 72
Bhabha, Homi 55
Black Skin, White Masks 40, 42
blanket protest 154, 155
Bloch, Ernst 82
body: assimilative theory of 38, 56, 62; as community 153–154; and food 8–9, 11; and mind, self-isolation through 83–89
body/mind: dualisms 15–17, 78–79; interconnectedness between 117–118, 120
body theories 12–18, 37–38, 44, 56, 93; *see also* closed body; open body

Bordo, Susan 3, 4, 83, 113
Bourdieu, Pierre 9–10, 12, 38, 56–57, 58
Boyle, Phelim P. 134
Brehon legal codes 151
Brillat-Savarin, Jean Amthelme 20
Butler, Judith 82

Carnal Appetites 5, 11
Cartesian dualism 16, 45, 69–70, 83, 92, 113–115, 157–158
Cartesianism 35
Cartesian model of the body 164
Cavanaugh, Jillian R. 43
Celts 161
Chapman, Michael 71, 72
civilizing mission 16
climatic disasters 47, 49; *see also* famine; Great Famine
closed body 13–16, 37–38, 44, 56, 57, 80, 83
Clough, Patricia 14
Coetzee, J. M. 2, 3, 26, 68–74, 77, 83, 91; *see also Life & Times of Michael K*
colonial and patriarchal power, female body in the intersections of 102, 105–107, 109–110, 116, 118
colonial development and education 123
colonial ideology, and hunger strike 38, 42, 44, 54
colonial/native bodies 21, 41
colonial Other, animalism of 41, 59
colonial/postcolonial body 17
colonial/postcolonial identity, politics of 37, 38, 44
colonial/postcolonial subject, transformation into 38–39
colonized body: 'animal-like' nature of the 88; theories of 40
criminal body 154

culinary choices, and colonization 57–58
Cultural Materialist framing 2
culture, impact of individual colonialist–native interaction on 107–108

Dangarembga, Tsitsi 2, 26, 104–105, 107, 108, 110, 127n59; *see also Nervous Conditions*
Davis, Mike 47
death-worlds 52
'Deciphering a Meal' 6–7
Deleuze, Gilles 81
Desai, Kiran 2, 11, 35, 43, 47, 54, 60; *see also Inheritance of Loss, The*
Descartes, René 16
desire and hunger relationship 89
Devereux, Stephen 48
dinner table 16, 42, 102, 106
Diplock Courts 149–150, 173n103
dirty protest 174n126
Distinction 9, 10, 58
Dorian, Hugh 172n58
Douglas, Mary 6

Eagleton, Terry 142, 160
eating: as a biotechnology 76; cultural process 35; dynamic process 35–36; and food values, ideological patterns of 36, 42, 44; and knowledge, links between 80–81; as a means of interpolating subjects 42; and 'Othering' process 36; and prisoners' bodies 156–157, 159
economics and gender inequality 105, 121
Ellmann, Maud 3, 9, 29n41, 42, 80, 116, 118, 159, 164
English, Richard 144
Essay on the Principle of Population 48

family meal 42
famine 6, 47, 49, 64–65n75; as an impact of British colonialism in India 51–54; and colonial administration failure 47; and food insecurity 1, 4, 7, 48; and food shortage 49; historical studies about 46; and literature 46; and overpopulation 48; and social sciences 46; *see also* Great Famine
Famine in Zimbabwe 104
Famine: Social Crisis and Historical Change 46
Fanon, Frantz 21, 22, 40, 42
Feldman, Alan 153, 154
female, dietary hardship of 105

female colonial body, and identity 118–120
FitzGerald, Garret 146
Flight to Objectivity, The: Essays on Cartesianism and Culture 3
food: choice 9; colonial ideologies of 37, 42, 44, 55; deprivation 1, 36; ingestion, ideology of 56–57; as a language of colonialism 7; and language relationships 42–45, 81–82; as a notion of power 46–47, 51, 54; and power inequities 109; significations 149; and words/language, relationship between 115–116
Food and Culture: A Reader 4, 5
food-entitlement approach 48, 103–104
food insecurity 1, 2, 4, 5–7, 46; and colonial domination 24–26; contemporary models of 48; issues in South Africa 75; and racism 75–76
food refusal 42, 44, 45, 54, 76; and self-harm 117; and theories of western pathology 110–111
food security: impact of capitalist free market on 104; and NAD 75
food shortage 48
food studies/gastrocriticism 1, 4–5, 10
force-feeding 42–43, 44
Formations of Violence 153
Foucault, Michel 8, 11–12, 52, 81, 85, 88, 158
freedom, and food abnegation 79–80
French aesthetics and cultural practice 9–10

Gaelic culture 139, 144, 152, 160
gastrocritical approaches 1
global dining practices 19
globalization: and modernity 36; through food 10–11
globalized economy and inequality 53
global market 50–61
global overconsumption 48
Gordon, Richard 119
Great Famine 133–134; feminization of 166; and hunger sufferings 140–141; losses produced by 135; Nationalists' account on 135–137, 138, 139, 148; politization of 138; Revisionists' account on 135, 136; as a traumatic event 134–135, 138, 140, 167; *see also* Ireland; Sands, Bobby
Great Hunger, The 135, 136
Great Transformation, The 51
Guattari, Félix 81

habitus 9
Head, Dominic 70
Heathcliff and the Great Hunger: Studies in Irish Culture 142
homestead: and gender inequality 106–107; patriarchal traditionalism of 106
How to Dine, Dinners, and Dining 19
Hulme, John 146
humoral theories 20, 21
hunger 64–65n75; and biopower 76; and famine 47–49, 103; genealogy of 80; and overpopulation 47; as the primary driving force 77–78; as protest 1; in South African history 75–76
Hunger 53
Hunger Artists, The: Starving, Writing and Imprisonment 3, 10, 42, 80
hunger denied 120
'Hunger Power: The Embodied Protest of the Political Hunger Strike' 44
hunger strikers 36, 152; closed/Cartesian body model and 37–38, 44; control over their bodies 17; *see also* Sands, Bobby
hunger strikes 110–112, 153; alimentary logic of 37; as anti-colonial struggles 2; and body theories 13–14; and Cartesian duality, Michael K's 70–71; and Cartesian model 87; connection made between food and language in 37, 44; contradictions of the colonial 22; as a discursive power of the body 45; and historical hunger 22–25; and necropolitical control 88–89; as ontological resistances 3; and postcolonial hunger 12; as a primitive forms of protest 45; as protest 155; racial and alimentary politics in 37; and self-immolation 44; as a somatic practice 15; structure of 37; subjectification model of 85
hunger-striking mechanism 22–23
hungry readers 89–95

identity and food, link between 9–10, 44, 55–56
Iliffe, John 104
Inheritance of Loss, The 2, 11, 35, 43, 47, 51; expressions of hunger in 59; history of hunger 46–54; hunger striker 36–46; narrative representations in 54–62
'invisible hand' 50

Ireland: anti-colonial discourses 143–144; blanket protest 154, 155; and British colonialism 142–143, 146; and British ideology 142; Celtic ritual practices 152; citizenship and land 144; cottiers 133; depictions of the Irish subject 151; and England 162–163; English–Irish dualism 145–146; factions within 148; and feminization of Irish body 162–163; hunger strike 1981 151–155; hunger strikes in 152; Manichean principles Nationalism 145; martyrdom and self-sacrifice in 151–152; masculine nationalities 163; modernity 148–149, 164, 165; Nationalism 136, 139–140, 142, 143–144, 154, 160, 161, 162, 164; open body models 152, 153–154; Poor Law of 1847 136; potato cultivation 134; poverty in 133; public consciousness 143; and race 161–162; Republicanism 144–145, 163, 164; Revisionism 146–147; romanticism 144–145; starving bodies of peasantry 141–142; *troscad* form 152–153; Troubles 2, 27, 31, 144, 146, 148, 149, 154, 163, 167, 169n1; *see also* Great Famine
Irish Republican Army (IRA) 26, 131, 146, 147; and hunger strike 151, 163–164; hunger strike 1981 155–169; hunger strikers' demands 154; inhuman conditions in prison 149; Irish militant versus cultural nationalisms 147–148; and motives behind hunger strikes 153; and Nationalisms 149; surveillance and cavity search 155–156; teleological vision of history 147; and traumatic hunger 150

Jameson, Frederick 74, 82
J.M. Coetzee and the Ethics of Reading 90

Kelleher, Margaret 46, 135, 166
Kiberd, Declan 142
Kinealy, Christine 141
knowledge and eating, links between 80–81

laissez-faire economy 24, 50, 52
language of food and hunger 107, 115–116
Last Conquest of Ireland (Perhaps), The 135–136
Late Victorian Holocausts 47
Levi, Primo 137

Life & Times of Michael K 68–95; ahistoricism of 69–72; allegory of 72–74; body and mind 83–89; and hungry readers 89–95; narrative and hunger 76–83
Lionnet, Françoise 1
Lloyd, David 135, 137, 138, 140, 147, 151
Location of Culture, The 55
Long Kesh prison 145, 151, 155, 156; *see also* Sands, Bobby
Loomba, Ania 3
Lupton, Deborah 42

Machin, Amanda 44
Mac Suibhne, Breandán 137
MacSwiney, Terence 152
malnutrition, and starvation 108
Malthusian approaches 48, 50–53, 103
Manichean principles 145, 161, 163, 164
Marxism and Form 82
Maxwell, Daniel 49
Mayer, Tamar 163
Mbembe, Achille 24, 52, 88
McKittrick, David 146
McVea, David 146
Meuret, Isabelle 6
Minor Transnationalism 1
Mintz, Sidney 21
Mitchel, John 135, 144
modernity, and globalization 36
Mokyr, Joel 134
Moody, Alys 71, 74
Morash, Chris 139

Native Affairs Department (NAD) 75
native body 17, 21–22, 40, 69, 109, 160
necropolitics 24, 52, 88, 135, 150
'Necropolitics' 52
Nervous Conditions 2; colonial domination and hunger in 102; female bodies in 109–112; food refusal and theories of western pathology in 110–111; narrative of 101–102; reasons for food dearth in 104; Rhodesia and hunger in 103–120
New Historicist framing 2
Ó Gráda, Cormac 134, 170n10

O'Loughlin, Michael 140
open body 13–16, 37–38, 44, 56, 57, 94–95, 152
Othered colonial body 78–79

Parry, Benita 143
Pearse, Patrick 147
Peel, Robert 137
permeable body *see* open body
Pfeifer, Michelle 23
Plato 15
polarized hierarchy 58
political body 95, 154, 163
Political Economy of Underdevelopment, The 53
Polyani, Karl 51
'population checks' 103
postcolonial hungers 2, 38, 54–55
post-Enlightenment Cartesian body 113, 119
potato famine 141
Poverty and Famines 48
power: alimentary hierarchy of 53–54; colonizer's discourse, biopolitics of 44; permutations of 61–62
prisoner's body 154, 155; *see also* Irish Republican Army (IRA); Sands, Bobby
Probyn, Elspeth 5, 12, 94
purchasing power, of individuals 48

Quilligan, Maureen 90

Rabinow, Paul 85
race theories: and colonial/postcolonial body 17–22; of native bodies 41, 45
reading and eating, connection between 114–115, 123
realism and radicalism 71–72
representation and materiality, dynamics between 70–72
restaurants 4, 10, 58–59, 60
Riley, Kathleen C.
Robert, Malthus T. 48, 50
Robertson, William 17
Roy, Parama 51

Said, Edward 20
Sands, Bobby 2, 24, 131, 140–141; 'Alone and Condemned' 159; on British domination and occupation 157; and Cartesianism 169; 'Castlereagh Trilogy, The' 150–151; death of 167–168; definition and meaning of body 160–161; on English–Irish dualism 145–146, 148, 154; 'Fenian Vermin, Etc.' 149; on Great Famine 173n96; hunger, as traumatic history reassertion 165–166; hunger references 155; and hunger strike 155–169; 'Modern Times' 148,

149; motivations for hunger strikes 168; narrativization of famine trauma 141; *One Day in My Life* 155–156, 157; on prison food 157; prison writings 132–133, 141, 142–151, 159, 172n77; 'Privileged Effort, The' 147; 'Rhyme of Time, The' 158; starving body of 141–142, 164, 165; 'Things Remain the Same – Torturous' 145; 'Trilogy, Diplock Court' 149–150; voracious reading by 159–160; 'We Won't be Fooled' 147; 'Window of My Mind, The' 145–146
Scarry, Elaine 159, 168
Sedgewick, Eve 14
self-abnegation 37, 38, 42, 45, 46, 84
self-cannibalization 59
self-immolation and hunger strike 44
self-starvation, sociohistorical meanings of 112–113
Sen, Amartya 48, 103
Shih, Shu-mei 1
silence, and the starving body 85
Sinn Féin 167
Slemon, Stephen 72, 74
Smith, Adam 17, 50
Social Democratic and Labour Party (SDLP) 173n85
somatic ontology 13, 35
South Africa, food equality issues in 75
Spielman, David 43, 44
stadial theory 17, 19, 20, 21

starvation 15, 17, 46, 64–65n75, 75, 108, 148
Sweetness and Power 21

taste, hierarchical system of 58
taste theory (Bourdieu) 12
Thatcher, Margaret 154, 163
Theories of Famine 48
Tóibín, Colm 136
Toward a Psychosociology of Contemporary Food Consumption 7
Trevelyan, Sir Charles 136

Unbearable Weight: Feminism, Western Culture and the Body 3, 113
Unilateral Declaration of Independence (UDI) 125n1

Vaughan, Michael 71
Vernon, James 53
Vital, Anthony 70, 71

Watson, Stephen 82
Wheeler, Roxanne 17
white bodies 21
Wilk, Richard 10
women's hunger 105–106
Woodham-Smith, Cecil 135, 136, 142

Yeats, William Butler 147

Zwicker, Heather 112

For Product Safety Concerns and Information please contact our EU representative GPSR@taylorandfrancis.com Taylor & Francis Verlag GmbH, Kaufingerstraße 24, 80331 München, Germany

Printed and bound by CPI Group (UK) Ltd, Croydon, CR0 4YY
08/06/2025
01897009-0005